T0075940

Introduction to Mathematical Modeling and Computer Simulations

Introduction to Mathematical Modeling and Computer Simulations

Vladimir Mityushev
Wojciech Nawalaniec
Natalia Rylko

CRC Press is an imprint of the
Taylor & Francis Group, an **informa** business
A CHAPMAN & HALL BOOK

CRC Press
Taylor & Francis Group
6000 Broken Sound Parkway NW, Suite 300
Boca Raton, FL 33487-2742

Printed on acid-free paper
Version Date: 20180112

International Standard Book Number-13: 978-1-138-19765-7 (Hardback)

Visit the Taylor & Francis Web site at
http://www.taylorandfrancis.com

and the CRC Press Web site at
http://www.crcpress.com

Printed and bound in Great Britain by
TJ International Ltd, Padstow, Cornwall

Contents

List of Figures ix

List of Tables xiii

Preface xv

I General Principles and Methods 1

1 Principles of Mathematical Modeling 3

1.1 How to develop a mathematical model 4
1.1.1 A simple mathematical model 4
1.1.2 Use of a computer 6
1.1.3 Development of mathematical models 12
1.2 Types of models 16
1.3 Stability of models 21
1.4 Dimension, units, and scaling 22
1.4.1 Dimensional analysis 22
1.4.2 Scaling 24
Exercises 25

2 Numerical and symbolic computations 27

2.1 Numerical and symbolic computations of derivatives and integrals 27
2.2 Iterative methods 29
2.3 Newton's method 30
2.4 Method of successive approximations 32
2.5 Banach Fixed Point Theorem 34
2.6 Why is it difficult to numerically solve some equations? . . . 37
Exercises 40

II Basic Applications 43

3 Application of calculus to classic mechanics **45**

3.1 Mechanical meaning of the derivative 45
3.2 Interpolation 46
3.3 Integrals 52
3.4 Potential energy 54
Exercises 55

4 Ordinary differential equations and their applications **57**

4.1 Principle of transition for ODE 58
4.2 Radioactive decay 59
4.3 Logistic differential equation and its modifications 61
4.3.1 Logistic differential equation 61
4.3.2 Modified logistic equation 61
4.3.3 Stability analysis 64
4.3.4 Bifurcation 66
4.4 Time delay 68
4.5 Approximate solution to differential equations 68
4.5.1 Taylor approximations 69
4.5.2 Padé approximations 71
4.6 Harmonic oscillation 73
4.6.1 Simple harmonic motion 73
4.6.2 Harmonic oscillator with friction and exterior forces . 76
4.6.3 Resonance 80
4.7 Lotka-Volterra model 83
4.8 Linearization 86
Exercises 88

5 Stochastic models **93**

5.1 Method of least squares 93
5.2 Fitting 97
5.3 Method of Monte Carlo 102
5.4 Random walk 105
Exercises 109

6 One-dimensional stationary problems **113**

6.1 1D geometry 113
6.2 Second order equations 116
6.3 1D Green's function 120
6.4 Green's function as a source 123
6.5 The δ–function 126
Exercises 130

III Advanced Applications 131

7 Vector analysis 133

7.1 Euclidean space $\mathbb{R}^3$ 133
7.1.1 Polar coordinates 135
7.1.2 Cylindrical coordinates 135
7.1.3 Spherical coordinates 137
7.2 Scalar, vector and mixed products 138
7.3 Rotation of bodies 142
7.4 Scalar, vector and mixed product in *Mathematica* 144
7.5 Tensors 145
7.6 Scalar and vector fields 148
7.6.1 Gradient 148
7.6.2 Divergence 152
7.6.3 Curl 156
7.6.4 Formulae for gradient, divergence and curl 156
7.7 Integral theorems 159
Exercises 161

8 Heat equations 163

8.1 Heat conduction equations 163
8.2 Initial and boundary value problems 168
8.3 Green's function for the 1D heat equation 169
8.4 Fourier series 173
8.5 Separation of variables 177
8.6 Discrete approximations of PDE 180
8.6.1 Finite-difference method 180
8.6.2 1D finite element method 183
8.6.3 Finite element method in $\mathbb{R}^2$ 186
8.7 Universality in Mathematical Modeling. Table 189
Exercises 189

9 Asymptotic methods in composites 193

9.1 Effective properties of composites 194
9.1.1 General introduction 194
9.1.2 Strategy of investigations 196
9.2 Maxwell's approach 199
9.2.1 Single-inclusion problem 199
9.2.2 Self consistent approximation 201
9.3 Densely packed balls 202
9.3.1 Cubic array 202
9.3.2 Densely packed balls and Voronoi diagrams 204

9.3.3 Optimal random packing 208
Exercises 213

Bibliography **217**

Index **221**

List of Figures

1.1 Falling of material object (point) 4
1.2 The expected trajectory (solid line) and its bounds (dashed line). 17
1.3 Radius $S(x)$ of the circular section at the point x. 18
1.4 Partition of the cylinder. The total mass of the cylinder is approximated by the sum of the masses of small cylinders. 18
1.5 Illustration of the black box. A set of inputs and outputs determines the object without any study of the interior structure. 20
1.6 Stable (a) and not stable (b) locations of a ball 21

2.1 Take a zero approximation x_0 and construct the point $M_0(x_0, f(x_0))$. The tangent line at this point to the graph of a function $y = f(x)$ passes through the axis OX in the point x_1 considered as the first approximation. In accordance with the geometric interpretation of the derivative we have $f'(x_0) = \tan\alpha$. On the other hand, we have $\tan\alpha = \frac{f(x_0)}{x_1 - x_0}$ by consideration of the triangle $M_0 x_0 x_1$. Hence, $f'(x_0) = \frac{f(x_0)}{x_1 - x_0}$ that yields equation (2.7). A point x_2 is constructed in the same way through x_1 and so on. In the present example, the iterations converge to a root of equation (2.5). 31
2.2 Successive approximations $x_1, x_2, \ldots$ are constructed in accordance with the rule described in Fig.2.1. In this example, the sequence $\{x_k\}$ diverges. 32
2.3 Graph of $P_{50}(x)$. 38
2.4 Graph of $y(x) = 5\cos x - x$ 39

4.1 Graph of (4.22) for $\alpha = 1$ and $\beta = 0.3$ (solid line), $\beta = 1$ (dashed line) and $\beta = 2$ (dotted line). 62
4.2 Graphs of the functions $f(u, 0.3, 20)$ (solid line) and $f(u, 1, 15)$ (dashed line). One can analyze the roots of the functions for small u on the right figure. Application of **DSolve** gives three positive roots $u_1 = 0.326574$, $u_2 = 3.87692$, $u_3 = 15.7965$ for equation $f(u, 0.3, 20) = 0$ and one root $u_1 = 13.9286$ for equation $f(u, 1, 15) = 0$. 65
4.3 Graphs of $x(t) = t^2 - \varepsilon$ for $\varepsilon = 0$ (solid line), for $\varepsilon > 0$ (dashed line) and for $\varepsilon < 0$ (dotted line). 66

4.4 Double root curve (4.32). Equation (4.30) has three positive roots for q and r in the domain D_3 and one root in the domain D_1. . 67
4.5 A material point and massless spring attached to the celling . . 74
4.6 Phase space ellipse (4.44) with a fixed energy E. 74
4.7 The phase portrait of equations (4.58), (4.60) for $a = 1.5$, $k = 6$, $b = 1$, $\ell = 4$. The stationary point $Q = (4, 4)$. 85

5.1 Line near a set of points 94
5.2 Square filled with the cloud of random points. 103
5.3 2D Brownian motion. a) The original location (disk) is connected to the ultimate location (square) after 1000 random steps. b) 10000 random steps. The linear size is 10 times less than in the figure (a). 108

6.1 1D temperature distribution in the 3D wall. 113
6.2 Let a function $u(x,t)$ is defined in the half-plane $\{(x, y, z) \in \mathbb{R}^3 : x > 0\}$ and does not depend on y and z. Then, the function $u(x,t)$ can be investigated as a function in the 1D space of variable x on the half-axis $x > 0$. 114
6.3 Let the temperature distribution be same in every section perpendicular to the cylinder axis. Then instead of the cylinder in $\mathbb{R}^3$ we can consider a disk in $\mathbb{R}^2$. Moreover, if the temperature distribution in every point of the disk depends only on the distance form the point O, then instead of the disk in $\mathbb{R}^2$ we can consider an interval $0 \leq r \leq r_0$, where r_0 denotes the radius of the disk. 115
6.4 Let the temperature distribution in the ball of the radius r_0 be dependent only on the distance r to the center of the ball. Then, the temperature distribution depends only on $r \in [0, r_0]$. 115
6.5 Local presentation of the domain as a half-plane in the vicinity of the point x_0. 116
6.6 Linear temperature distribution u in the interval $[a, b]$. 117
6.7 The temperature distribution u, the heat flux $q = -\lambda u'$ and the functions $u''(x) = -H(x - x_0)$. The data are for $\lambda = 1$. 120
6.8 Graph of $g(x, \xi)$ 123
6.9 Graph of the function u given by (6.43). 127
6.10 Graphs of $\Phi_m(x)$ for $m = 1, 2, \ldots, 5$ where $\Phi_1(x) = \frac{1}{\sqrt{\pi}} e^{-x^2}$. . . 128

7.1 Polar coordinate system. 135
7.2 Cylindrical coordinate system 136
7.3 Fibrous composite. 136
7.4 Spherical coordinate system. 137
7.5 Geometrical interpretation of the scalar product. 138
7.6 Force $\mathbf{F}$ and its decomposition. 138
7.7 To item iii) of Definition 7.2: $\odot$ means that the vector $\mathbf{c} = \mathbf{a} \times \mathbf{b}$ is directed to the Reader; $\otimes$ from the Reader. 139
7.8 Geometric interpretation of formula (7.23). 140

7.9 Tetrahedron and the vectors $\mathbf{S}_i$. 142
7.10 Rotation of the point M about the x_3-axis with the angle velocity ω. 143
7.11 Illustration of the natural parametrization. 149
7.12 The level set of the function $u(x_1, x_2) = x_1^2 + x_2^2$ are circles $x_1^2 + x_2^2 = c$. 150
7.13 The unit tangent and normal vectors, $\mathbf{s}$ and $\mathbf{n}$, respectively, to L at the point $\mathbf{x}$. The direction of greatest increase of $u = u(x_1, x_2)$ at $\mathbf{x}$ coincides with ∇u. In the considered case, it is $\mathbf{n}$; the direction of greatest decrease is $(-\mathbf{n})$. The function does not change in the direction $\pm\mathbf{s}$. 151
7.14 The orthogonal nets generated by the function $u(x_1, x_2) = x_1^2 + x_2^2$. 152
7.15 Singular and laminar fields. 153

8.1 Heat flux in the x_1-direction. 165
8.2 Graph of the function f. 174
8.3 Graphs of the functions $y = f_1(x, n)$ dla $n = 5, 20, 50, 150$. 175
8.4 Mesh points with the double numeration (i, j); layers $j = 0$, $j = 1$, $j = 2$ etc. 182
8.5 Plots of the basic functions. 184
8.6 a) Proper triangulation; b) non-allowable triangulation. 187
8.7 Basic function $u_j(\mathbf{x})$ is a linear function in each triangle Δ_m, equal to 1 at the central point $\mathbf{v}_j$ and 0 at other vertices. 188

9.1 Spherical inclusions embedded in host. 195
9.2 Methodologically wrong scheme of the RVE from WikipediA *en.wikipedia.org/wiki/Representative_elementary_volume*. The random (on the left) and regular (on the right) composites in Fig.9.2 can have significantly different macroscopic properties in the case of percolation displayed in Fig.9.5. 197
9.3 2D interpretation.: square with random points forms a Delaunay graph. 198
9.4 a) Cubic array; b) its cross section near the gap between two balls. 202
9.5 2D periodic discrete network with two percolation chains connecting the opposite sides of the unit cell. 206
9.6 2D periodic discrete network with points instead of balls. Compare to the bounded discrete network displayed in Fig.9.3. 206
9.7 a) Three points in the cell Q_0 are distinguished. Dashed lines show the lattice, solid lines the double periodic Delaunay graph. b) The optimal graph isomorphic to the graph from a). 213

List of Tables

5.1 Data table. 93
5.2 Description of the random value X. 105

8.1 Universality in Mathematical Modeling [10]. Physical phenomena and corresponding vector fields $\mathbf{j}(\mathbf{x})$, $\mathbf{E}(\mathbf{x})$; tensors $\boldsymbol{\sigma}(\mathbf{x})$ and scalars for anisotropic and isotropic media, respectively. The fields satisfy equations $\mathbf{j}(\mathbf{x}) = \boldsymbol{\sigma}(\mathbf{x})\mathbf{e}(\mathbf{x})$ and $\nabla \cdot \mathbf{j} = 0$. In 1D case, the operator ∇ is the derivative, i.e., $\nabla = \frac{d}{dx}$. Then, $j(x)$, $E(x)$ and $\sigma(x)$ are scalar function in x. Equations $j(x) = \sigma(x)e(x)$ and $\frac{dj}{dx} = 0$ are fulfilled for 1D media. 189

Preface

"Philosophy is written in that great book which ever lies before our eyes - I mean the universe - but we cannot understand it if we do not first learn the language and grasp the symbols, in which it is written. This book is written in the mathematical language, and the symbols are triangles, circles and other geometrical figures, without whose help it is impossible to comprehend a single word of it; without which one wanders in vain through a dark labyrinth"

Galileo Galilei

To learn or not to learn mathematics? Such Shakespearean questions are posed by engineers. Perhaps, to learn linear algebraic systems for numerical methods but rather not to learn advanced integrals because they can be calculated by computer. A mathematician can pay too much attention to existence of the integral. For an engineer, the existence of the integral means for instance, existence of the mass of a given body. One can imagine the engineer's opinion about a mathematician who discusses an object without mass. An engineer can think that "a mathematician discusses strange surface integrals instead of the integrals needed to calculate the heat flux through the surfaces". However, it turns out that the engineer is interested in the same integrals[1]. Now, one can imagine what a mathematician thinks about an engineer. This book is written, in particular, to avoid such misleading declarations from both sides.

Development of new technologies, new materials and technological objects requires theoretical mathematical descriptions of new objects and study of their behavior. Engineers then have to make frequently expensive and long experiments, which are sometimes dangerous. It is possible to gather a small set of a data and further to extend the results by the use of Mathematical Modeling. So, mathematical and computer calculations can be considered as a continuation of the physical experiments[2]. Consider, for instance, creation of new composites. Let an engineer note that a new substance added to a material makes it better for certain purposes. A problem of describing this phenomenon then arises. Does the temperature impact on new material? What is the optimal concentration of the admixture? Many answers can be obtained not by hard experimental works, but by Mathematical Modeling. Let a new composite consist of 10 different substances which can be embedded in

[1] it is a true story

[2] V.I. Arnold [8]: "Mathematics is the part of physics where experiments are cheap."

material with various concentrations. In order to optimize the properties of the composite let an engineer try the concentrations 7%, 10% and 13% of each component. Then the experimental problem to determine the optimal composite is reduced to the investigation of $3^9 = 19683$ samples. At the same time one can use the known formulae for the effective properties and quickly obtain the results.

This textbook is intended for readers who want to understand the main principles of Modeling and Simulations in settings that are important for the applications without using profound mathematical tools required by most advanced texts. It can be useful for beginning applied mathematicians and engineers who use Mathematical Modeling. Our goal is to outline Mathematical Modeling using simple mathematical description that make it accessible for first- and second-year students (undergraduate courses for bachelor's degrees). This book consists of three parts. Part I "General Principles and Methods" is an elementary introduction to Mathematical Modeling based on the introductory mathematical courses. We think that this is the main part which should be worked out by a beginner in Applied Mathematics and other sciences addressed to Mathematical Modeling. There are general principles, methods and tricks of Mathematical Modeling used in different sciences. For instance, it is useful to introduce a linear operator in general form and later to describe linear models of econometrics. The didactic principle of primary introduction of examples and further of a theory narrows down applications of Mathematical Modeling. In this book, the linear operator is not rigorously introduced in a general space but it is outlined for further applications. In econometrics, it is better first to study the Method of Least Squares as minimizing a quadratic function and after to learn economical terms. In engineering, first it is preferable to study the vector calculus and after to describe topics from fluid and solid mechanics. Such an approach facilitates understanding special topics. Universality of Mathematical Modeling and the corresponding equations helps to understand such formally different phenomena as the behavior of solids and liquids for engineers, credit and debit for economists, etc.

Part II contains the fundamental mathematical methods and examples used in Mathematical Modeling. Parts I and II are written in such a simple form that it can be presented to second-year students at the undergraduate level familiar with calculus and elementary ordinary differential equations. I suppose that Parts I and II can be considered as an introductory course to general problems of Mathematical Modeling, in particular, to Industrial Mathematics. Other examples from engineering, economics, computer sciences, biology are widely presented in the books [3], [13], [31], [32], [33], [38], [39], [40], [43], [48], [49], [61].

Part III concerns vector calculus. The main attention is paid to computer implementation of the main calculus operators. Further, initial and boundary value problems are discussed for the heat equation. This part can be considered as an example of constructive methods applied to ordinary and partial differential equations suitable for students at the graduate level. There are many nice textbooks devoted to such problems [16], [31], [41], [52] where the fundamental mathematical methods such as separation of variables, discrete and integral transforms etc., can be combined with the computer approach presented here.

The main point of this book is thewide use of computer for numerical and symbolic computations, and for graphic presentations. We suggest using the computer while reading this book, especially the one with the package *Mathematica*® (see courses in the books [60, 56, 28, 30, 43, 27]). Other software packages can be used but special codes should be prepared then. One may read this book without using any computer code, but many features of Mathematical Modeling could be lost. So, it is better to pay one hour to *Mathematica* to apply at least simple operators and completely use the material presented in this book. Use of *Mathematica* is caused by its possibilities to apply symbolic computations. Frequently, methods of Applied Mathematics are reduced by users to numerical packages without addressing to the corresponding mathematical description that can lead to missing of the important features of the considered real-life problem. We use symbolic computations to demonstrate advanced possibilities of computer simulations. Other alternatives to *Mathematica* are available. We use also MATLAB and insert in the book MATLAB-boxes containing codes and necessary descriptions [19, 29].

Besides necessary formal theoretical presentation, we try to explain "how to do it" revealing secrets and tricks used in practical applications. Informal advice presented as "principles" facilitates understanding the interdisciplinary approaches applied to real-life problems. An important feature of this book is the inclusion of exercises considered as projects with computer implementation. A lot of interesting projects with codes are selected at *http:// demonstrations.wolfram.com/*. They can be also used for a course on Mathematical Modeling and Computer Simulations.

Partially, this book is based on the textbook [46] written in Polish (open access at *http://mityu.up.krakow.pl/publication/*).

Kraków, *Vladimir Mityushev, Natalia Rylko and Wojciech Nawalaniec*
December 2017

Part I

General Principles and Methods

Chapter 1

Principles of Mathematical Modeling

1.1	How to develop a mathematical model	3
	1.1.1 A simple mathematical model	4
	1.1.2 Use of a computer	6
	1.1.3 Development of mathematical models	12
1.2	Types of models	16
1.3	Stability of models	21
1.4	Dimension, units, and scaling	22
	1.4.1 Dimensional analysis	22
	1.4.2 Scaling	24
	Exercises	25

Mathematical Modeling describes a process and an object by use of the mathematical language. A process or an object is presented in a "pure form" in Mathematical Modeling when external perturbations disturbing the study are absent. Computer simulation is a natural continuation of the Mathematical Modeling. *Computer simulation* can be considered as a computer experiment which corresponds to an experiment in the real world. Such a treatment is rather related to numerical simulations. Symbolic simulations yield more than just an experiment. They can be considered as a transformation of a mathematical model by computer, since symbolic simulations keep parameters of the model in symbolic form that corresponds to a set of actual experiments. One can obtain numerical results as in actual experiments only after substitutions of the symbolic parameters with the numerical data. Therefore, symbolic simulations complete the mathematical model and embrace actual experiments.

Mathematical Modeling of stochastic processes is based on the probability theory, in particular, that leads to using of random walks, Monte Carlo methods and the standard statistics tools.

Symbolic simulations are usually realized in the form of solution to equations in one unknown, to a system of linear algebraic equations, both ordinary and partial differential equations (ODE and PDE). Discrete methods such as Finite Element Methods or Finite–Difference Methods for finding approximate solutions of ODE and PDE are usually used in the form of numerical simulations.

1.1 How to develop a mathematical model

1.1.1 A simple mathematical model

Consider a simple example of the free falling object from a height h using Mathematical Modeling. How to describe the trajectory of the object? The first questions which should be posed are the questions "where" and "when", i.e., we have to describe the space where the free fall is taking place and the time when it happens. Let gravity be the only force acting upon the falling object. In this case, only one axis is needed to describe the space. Let it be the axis OY shown in Fig.1.1. It is necessary to fix a unit segment on the axis which shows the direction and the line unit. In Fig.1.1, the axis OY is chosen to point in the downward direction. Meters can be chosen as the length unit on OY. However, dimensionless units are frequently used in Mathematical Modeling, which enables the result to be obtained in the easiest way. Dimensionless results are transformed into dimension form at the last stage. We will discuss this question later in Sec.1.4. Now, we are going to fix a dimensionless line unit. The time unit t is also fixed as a dimensionless unit. It is assumed that the object begins to fall at the initial time $t = 0$.

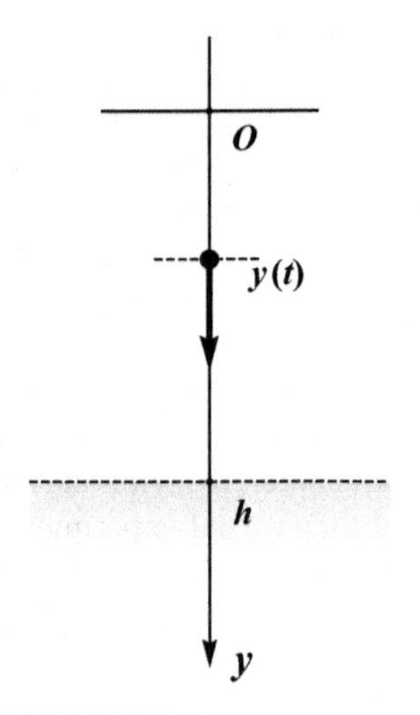

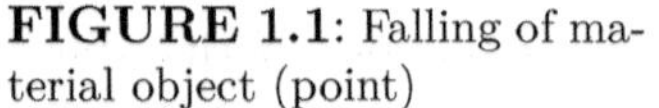

FIGURE 1.1: Falling of material object (point)

After answering the questions "where" and "when" we should analyse conditions of the fall. If the object is a stone, the air resistance can be ignored. However, it may not be done for a leaf. Let us study a stone now. Then one can consider the stone as a material point, i.e., as a point on the OY axis for each time t.

Next, one should use a physical law of the falling material point. It follows from the gravitational law that an object near the surface of the Earth falls with the constant acceleration g (equal to $9.81 \frac{m}{sek^2}$ in the SI units). *Projectile motion* is a motion in which a material point moves along a trajectory under the action of gravity only. Such a material point is called *projectile* .

Mathematics, "hard artillery", arises at the following step to formally describe reality. We are interested in representing the motion of the point in time as a function $y(t)$ which shows the coordinate of the material point in time t. The function $y(t)$ is called the trajectory in mechanics. The velocity of the point is the derivative with respect to time $v(t) = y'(t)$, acceleration is the second derivative $a(t) = y''(t)$. Now, one can write the law in the mathematical form

$$\text{“point falls with constant acceleration } g\text{”} \Leftrightarrow \quad a(t) = g \Leftrightarrow y''(t) = g \quad (1.1)$$

We have got the ordinary differential equation (1.1). Its general solution has the form

$$y(t) = \frac{gt^2}{2} + C_1 t + C_2, \tag{1.2}$$

where C_1 i C_2 are arbitrary constants. The solution (1.2) of equation (1.1) is obtained by double integration of (1.1), where C_1 and C_2 are constants of integrations. The trajectory of the material point has to be uniquely determined and could not depend on the arbitrary constants C_1 and C_2. How to fix them? First, the law (1.1) does not take into account the information about when and how the stone is thrown. In our case, the material point is located at the point $y = 0$ of the axis OY at the time $t = 0$. Hence,

$$y(0) = 0. \tag{1.3}$$

Moreover, the object can be thrown with any initial velocity v_0. If one just drops the stone, the initial velocity is equal to zero. If one applies a force and throws the stone down or up the initial velocity v_0 does not vanish. Hence, we have the second condition

$$y'(0) = v_0. \tag{1.4}$$

The conditions (1.3)–(1.4) are called initial in the theory of ordinary differential equations and the problem (1.3)–(1.4) is called Cauchy's problem. This problem is properly posed and has a unique solution. It is necessary to substitute $C_1 = v_0$ i $C_2 = 0$ in the general formula (1.2) in order to fulfil the initial conditions (1.3)–(1.4). As a result we obtain the required particular solution

$$y(t) = \frac{gt^2}{2} + v_0 t. \tag{1.5}$$

The mathematical problem has been solved. The height h has to be taken into account to find the time T when the projectile reaches the earth. The following equation with respect to T is obtained by substitution of $t = T$ and $y = h$ into (1.5)

$$h = \frac{gT^2}{2} + v_0 T. \tag{1.6}$$

Out of two solutions

$$T_1 = \frac{-v_0 + \sqrt{2gh + v_0^2}}{g}, \quad T_2 = \frac{-v_0 - \sqrt{2gh + v_0^2}}{g}$$

we must take the first one, since the second solution is always negative which contradicts the physical treatment of the problem ($T > 0$).

Remark 1.1. It is a common practice to precisely describe "operating conditions" in pure mathematics. For instance, in order to work with a function mathematicians must define a domain of the function. If a derivative is used, a mathematician must say "let a function $y(t)$ be differentiable in a set A". If a mathematician writes $\frac{a}{m}$, before this he/she must write $m \neq 0$. However, in applied mathematics the mass m is always non–negative. Thus, $\frac{a}{m}$ for $m = 0$ has no any sense. For instance, an applied mathematician writes $y'(t)$ assuming that this derivative exists. So, usually in applied mathematics if something is written, it is tacitly assumed that it exists.

1.1.2 Use of a computer

The above example has been solved "by hand" without any computer, since the calculations are simple. It is in accordance with the principle:

Principle of hand calculations. *If you can calculate something without any computer, try to do it "by hand". If you cannot or you are not sure in your calculations, try to use a computer.*

For instance, let somebody not be sure in equations (1.5)–(1.6) based on derivatives and integrals. Let us solve the problem by use of the operator **DSolve** in *Mathematica*[1]

```
In[1]:= DSolve[{x''[t] == g, x[0] == 0, x'[0] == v0},
          x[t], t]
```

$$\text{Out[1]= } \left\{\left\{x[t] \to \frac{1}{2}\left(g\,t^2 + 2\,t\,v0\right)\right\}\right\}$$

One can copy the resulting function and further paste it into the definition of the function $y(t)$. However, it is easier to extract $y(t)$ from the expression Out[1] as follows

```
In[2]:= y[t_] = %[[1, 1, 2]]
```

$$\text{Out[2]= } \frac{1}{2}\left(g\,t^2 + 2\,t\,v0\right)$$

Here, **%** means that the latest result is taken, i.e., Out[1], the triple **1,1,2** denotes the coordinates, so **1,1** yields the expression[2] **Rule[x[t],** $\frac{1}{2}(gt^2 + 2tv0)$**]**. Then, **2** means the second element out of this expression.

Solve the equation (1.6) and take the proper solution:

---

[1] In *Mathematica*, v_0 means the value of the function v at the point 0. We write the symbol $v0$ instead of v_0 in *Mathematica*'s code in order to use $v0$ as a separate independent variable.

[2] You can investigate full form the expressions with the **FullForm** operator.

```
In[3]:= Solve[h == y[t], t]
```

$$\text{Out[3]=}\ \left\{\left\{t \to \frac{-v0 - \sqrt{2\,g\,h + v0^2}}{g}\right\}, \left\{t \to \frac{-v0 + \sqrt{2\,g\,h + v0^2}}{g}\right\}\right\}$$

```
In[4]:= T = %[[2, 1, 2]]
```

$$\text{Out[4]=}\ \frac{-v0 + \sqrt{2\,g\,h + v0^2}}{g}$$

Fix the values of the parameters g, h, v_0 and prepare a graph to demonstrate the location of the material point in time. It can be done by use of the operator **Plot** as **Plot[h-y[t], {t,0,T}]**. To get a better visualisation one can do it in a more complicated way:

```
In[5]:= {g = 9.81, h = 0, v0 = -2};

In[6]:= Show[Plot[h - y[t], {t, 0, T}, AxesLabel → {"t", "z(t)"},
          LabelStyle → 16, PlotRange → {{-0.02, 0.45}, {-0.03, 0.25}},
          PlotStyle -> Directive[Darker[Gray, 0.1], Thick],
          AspectRatio → Automatic, TicksStyle → 12],
        Graphics[{Text[Style["T", FontSize → 15, Bold], {1.03 T, -0.015}],
          Text[Style["0", FontSize → 15, Bold], {0.03 T, -0.015}],
          {PointSize[Medium], Point /@ {{T, 0}, {0, 0}}}}],
        ImageSize → 450]
```

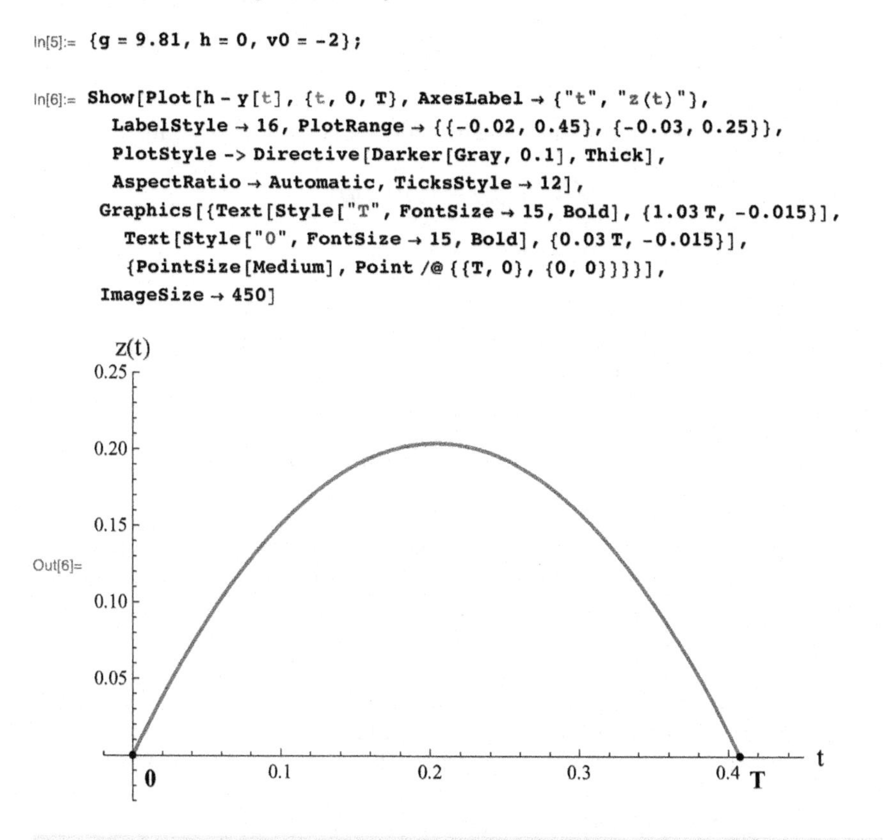

MATLAB Example Box 1.1

When computations are performed with the MATLAB package, one can use *Command Window* to enter commands at the command line (indicated by the prompt »). Let us define variable g

```
>> g  = 9.80665;
```

Now, we transform the considered equation (1.1) into a system of ODEs. Using $y_k' = y^{(k-1)}$ $(k = 1, 2, 3)$ we obtain

$$\begin{aligned} y_1'(t) &= y_2(t), \\ y_2'(t) &= y_3(t) = g, \end{aligned}$$

with the corresponding initial values

$$y_1(0) = h,$$
$$y_2(0) = v_0.$$

Let us define the initial values and place them into a single array

```
>> h  = -1;
>> v0 = -3;
>> init_vals = [h, v0]
```

Define the function representing right sides of the above system

$$\mathbf{f}(t, y_1, y_2) = (y_2(t), g) \tag{1.7}$$

using *anonymous function* as follows

```
>> f = @(t, y) [y(2); g];
```

We are going to solve the equation numerically, hence we have to provide the range of the integration

```
>> t_interval = [0 1];
```

Now we are ready to use the built-in solver function `ode45`

```
>> [T, Y] = ode45(f, t_interval ,init_vals);
```

Here, the solver returns two arrays: the array `T` contains time points corresponding to rows of the solution array `Y`. One can combine both arrays and plot the solution

```
>> plot(T, -Y(:,1))
```

The other way of computing in MATLAB, adapted in this book, is running scripts written in the MATLAB Editor (files with code organized by functions saved with `.m` extension). Below the source code of file `script01.m` is presented. Note that the very first function, which is called the *primary function*, should have the same name as the name of the file.

```
% primary function
function script01()

    % initial values
```

```
    v0 = -2;
    h  = -3;
    t_interval = [0 1];

    % solution of the ODE
    [T, Y] = ode45(@f, t_interval, [h v0]);

    % plot
    H = -Y(:,1);              % height of the body
    plot(T, H)
    xlabel('time')
    ylabel('height')
    grid on
    ylim([0 max(1.1*H)])

% subfunction
function dy = f(t, y)
    g  = 9.80665;
    dy = zeros(2,1);        % a column vector
    % yk denotes (k-1)st derivative of y
    dy(1) = y(2);           % y1'(t) = y2(t)
    dy(2) = g;              % y2'(t) = y3(t) = g
```

The script can be run simply by pressing the `Run` button from the Editor as well as by typing its name into the command window

```
>> script01
```

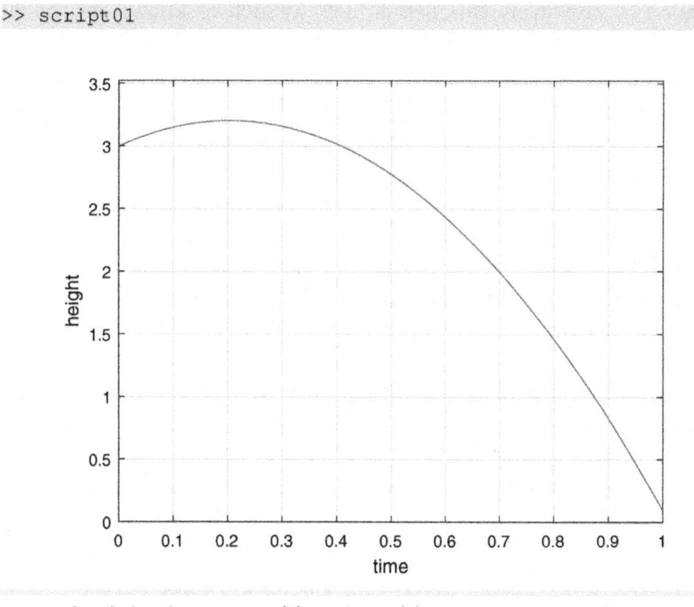

The graph of the function $z(t) = h - y(t)$ is drawn with all the numerically given parameters. After such a success one can think that an ordinary differential equation can be solved by means of computer without any course of calculus. It is enough just to write **`DSolve`** or something else in another package, to press a button and everything is displayed. But the question is, which buttons should be pressed? A computer is a continuation of our knowledge.

In order to perform simple computations and to solve simple problems a user need not know calculus. But a step aside from a standard way could lead to trouble. A poorly prepared user could rather repeat manipulations but could not solve a new simple problem. It is possible to give many examples when incorrect or superficial use of computer by engineers yields incorrect results. Below, we present such an example.

Calculate 10 partial sums of the series beginning with the partial sum 1160 for series

$$\sum_{k=1}^{\infty} \frac{1}{k} \tag{1.8}$$

We have[3]

```
In[1]:= s[n_] := Sum[1./k, {k, 1, n}]

In[2]:= Table[s[n], {n, 1660, 1670}]

Out[2]= {7.99209, 7.99269, 7.99329, 7.99389, 7.9945,
  7.9951, 7.9957, 7.9963, 7.9969, 7.9975, 7.99809}
```

Direct observations of the results could yield the conclusion that this series converges to 8. However, this series (1.8), which is known as the *harmonic series* diverges[4]. In order to avoid such mistakes one should use the following:

Principle of the stupid computer. *Do not trust the computer. If you do not know what are you computing, use the weaker principle of hand calculations. If it is possible to check the computer result "by hand", do it.*

However, the customer is not always right. A computer gives the correct answer to a question, but the user does not understand it and blames the computer on the basis of the above principle. Here, the following principle has to be taken into account:

Presumption principle of innocence of the computer. *The answer you get depends on the question you ask.*

Presumption principle of innocence of the computer completes the principle of the stupid computer. The computer answers the question and the computer does not care that the question is stupid. Its task is to answer the question in general, without comment, as in the army. It is worth mentioning that *Mathematica* sometimes warns us. Below, *Mathematica* warns about the divergent series

---

[3]The symbol **n_** means *for all n* ($\forall n$) in *Mathematica's* codes; **n_** or **x_** must be used for any argument in definitions of functions.

[4]it is possible to say in wider sense that it converges to $+\infty$

In[3]:= $\sum_{k=1}^{\infty} \frac{1}{k}$;

Sum::div : Sum does not converge. ≫

and division by zero

In[4]:= $\frac{17}{0}$

Power::infy : Infinite expression $\frac{1}{0}$ encountered. ≫

MATLAB Example Box 1.2

The following script plots partial sums of another divergent series

$$\sum_{k=2}^{\infty} \frac{1}{k \log(k)}. \tag{1.9}$$

```
function script02()

    % array of indexes [2,3,...,2000]
    k = [2:2000];

    % corresponding sequence and partial sums
    sequence = 1.0 ./ (k .* log(k));
    partial_sums = cumsum(sequence);

    % plot
    plot(k, partial_sums, '.k')
    grid on
```

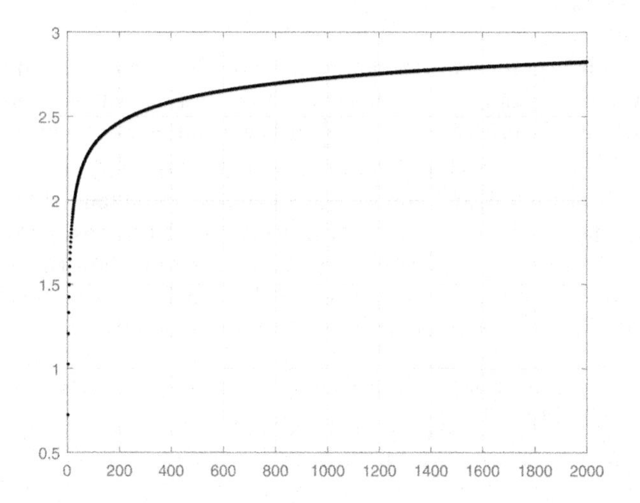

As we can see, such visual observation may also suggest convergence of the series. Note that the operators `./` and `.*` are examples of generalized arithmetic operations. Here, division and multiplications are performed element by element resulting in a new array. Here, the `cumsum` operator computes the cumulative sum (a vector of partial sums of a given vector).

1.1.3 Development of mathematical models

In Sec.1.1.1, steps of the development of a mathematical model are demonstrated in a simple example. These steps can be applied to general models.

Steps taken to create a mathematical model:

i) to introduce spatial variables (description of geometry) and time;
ii) to think about dimensional units; to introduce the units perhaps during solution of the mathematical problem;
iii) to introduce assumptions and conditions, first, as simple as possible;
iv) to formulate the law (physical, economical, biological, empirical etc.);
v) to develop a mathematical description, first, as simple as possible;
vi) to try to solve the mathematical problem "by hand"; if that does not work, to try a numerical method; to compare the results if different methods are applied;
vii) verification of the model; if the results are suspect to get back to the previous steps;

The following comment on the above steps is important.

Principle of the simplest model. *At the beginning do it in the simplest way, even it does not yield acceptable results.*[5]

In the step vi), it is worth doing most of the work "by hand" in order to simplify computations. We can say even more. For instance, instead of the computation of the series $S = \sum_{n=1}^{\infty} \frac{1}{n^2}$ it is worth looking up in the handbook (for example, implemented in *Mathematica*) and rewriting the result $S = \frac{\pi^2}{6}$.

Thus, it is not necessary to perform calculations "by hand"[6]. For instance, the best method to calculate an integral is to find the answer in the handbook[7]. Hence, the most important is to know where to find the answer. This requires good mathematical education. To know where to look is also a type of knowledge. In this case, another type of memory is required, not the one

[5]c.f. Albert Einstein: "Everything should be made as simple as possible, but not simpler."
[6]if only it is useful as an exercise
[7]Information as a part of knowledge becomes very important in the modern world. An apt example is *Google Search* used by students and researchers more often than printed textbooks

that is frequently required to pass exams. In particular, development of mathematical models means the proper selection of existing models scattered in textbooks, internet etc.

During development of the model one should mind that the model has to correspond to the real phenomena both qualitatively and quantitatively. This contradicts the principle of the simplest model. Anyway, we recommend following this principle gradually tending to be in correspondence with reality.[8]

After verification of the simple model, this model can be improved by advanced study. For instance, if instead of the stone (see Sec.1.1.1) a leaf is falling, it is better first to discuss the falling of the leaf in vacuum. Then, it is possible to complete the model taking into account air resistance. Following this approach we pass from the stone to the leaf below. The equation of the force equilibrium is similar to (1.1):

$$m\, y''(t) = m\, g, \tag{1.10}$$

where the exterior force $ma(t) = my''(t)$ presented in the left part in accordance with Newton's law. The gravitational force mg is in the right part. The corrected model for the leaf has the form

$$m\, y''(t) = m\, g - k\, v(t), \tag{1.11}$$

where $F = -kv(t)$ with $k \geq 0$ is the air resistance. It is assumed here that the air resistance is proportional to the velocity of the fall $v(t)$. It is an empirical law, sometimes it has the form $F = -kv^2(t)$ or may be more complicated. Then equation (1.11) becomes

$$y''(t) + \frac{k}{m}\, y'(t) = g. \tag{1.12}$$

The differential equation (1.12) with the initial conditions (1.3)–(1.4) has a unique solution.

```
In[1]:= DSolve[{y''[t] == g - k/m y'[t], y[0] == 0, y'[0] == v0}, y[t], t]
```

$$\text{Out[1]=}\ \left\{\left\{y[t] \to \frac{e^{-\frac{k t}{m}}\, m \left(g m - e^{\frac{k t}{m}} g m + e^{\frac{k t}{m}} g k t - k v_0 + e^{\frac{k t}{m}} k v_0\right)}{k^2}\right\}\right\}$$

```
In[2]:= %[[1, 1, 2]] // Expand
```

$$\text{Out[2]=}\ -\frac{g m^2}{k^2} + \frac{e^{-\frac{k t}{m}} g m^2}{k^2} + \frac{g m t}{k} + \frac{m v_0}{k} - \frac{e^{-\frac{k t}{m}} m v_0}{k}$$

The trajectory $y(t)$ has been constructed. Further investigation and computations can be performed by the scheme presented in the previous Secs 1.1.1-1.1.2.

---

[8]Zvi Artstein [9]: "Nature is a very good approximation of Mathematics".

MATLAB Example Box 1.3

One can simulate the considered model numerically as well and observe how the acceleration decreases due to the air resistance. Consider the following MATLAB script.

```
function script03()
    % parameters
    g  = 9.80665;
    v0 = 0;
    m  = 1;
    k  = 1;
    h  = 100;
    t_interval = [0 10];

    % solution of the equation
    f = @(t, y) f(t, y, v0, m, k, g);
    [T, X] = ode45(f, t_interval, [0 v0]);

    % plots of height and velocity in time
    H = h - X(:,1);              % height
    V = -X(:,2);                 % speed

    subplot(1,2,1); % left subplot
    plot(T, H)
    xlabel('time')
    ylabel('height')
    grid on

    subplot(1,2,2); % right subplot
    plot(T,V)
    xlabel('time')
    ylabel('speed')
    grid on

function dy = f(t, y, v0, m, k, g)
    dy = zeros(2,1);
    dy(1) = y(2);
    dy(2) = g - k/m * y(2);
```

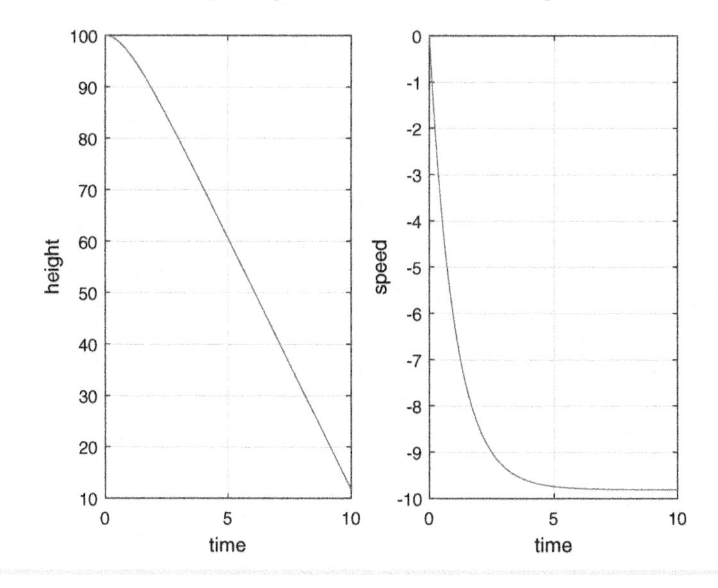

The final solution should not be given by exact formulae. Consider the example of the cooling coffee described in the excellent book by Amelkin [5].

Example 1.1. Anatol and Vladimir ordered two cups of coffee with cream in a bar. Two cups of hot coffee with the same temperature are simultaneously prepared. Anatol and Vladimir take the following steps. Anatol adds cream to coffee and covers the cup by a napkin. Vladimir covers the cup by a napkin and adds cream to his coffee 10 min later when they simultaneously begin to drink their coffee. Who drinks hotter coffee?

In order to solve this problem in accordance with the book [5] one can develop a mathematical model of the cooling coffee based on the theory of heat conduction, solve the stated problems and calculate the temperature of Anatol's and Vladimir's coffees 10 min later.

However, in this case we are interested only in the question, in which cup the coffee is hotter. We are not interested in the values of the temperature. The stated problem can be formulated in another way. When does coffee lose less heat in 10 min? Does it happen when Anatol decreases its temperature by adding cream at the time $t = 0$ min or when coffee is cooled at the time $t = 10$ min? In order to answer this question we need only one law of the heat conduction: *dissipation (loss) of heat is proportional to the difference of temperatures of the object (coffee) and environment.* Hence, hotter coffee loses more heat. Therefore, Anatol drinks hotter coffee than Vladimir.

Similar questions arise in other regimes of heating and cooling, for instance in buildings. In order to reach a given temperature at a fixed time with minimal dissipation one has to heat the object at that fixed time, not earlier.

The above problem is based on the estimations of energy. We do not take into account the daily change of prices of energy[9].

Principle of the energy estimations. *Estimate energy before you make a deal.*

This principle helps us to answer the following questions. Is the force of the artillery gun sufficient to overcome the gravitational force to go into space? (the answer is negative) Can the Moon cause an earthquake? "Does the flap of a butterfly's wings in Brazil set off a tornado in Texas?"[10] Edward N. Lorenz asked this question and illustrated it by an example of dynamical system which describes the *butterfly effect* when a small change in one state of a deterministic system can result in large differences in a later state. Such a question is closely related to stability discussed in Sec.1.3.

1.2 Types of models

Types of the mathematical models can be the following:

a) deterministic and stochastic (random);
b) discrete and continuous;
c) linear and non–linear.

a) An example of the deterministic model is given in Sec.1.1.1, see for instance equation (1.12). Let one check the model by means of experiments and see that a light wind impacts the result. But how to introduce this wind into the model? How to estimate the random influence of the wind? Equation (1.12) was useful but should be revised. Let us consider the question from another point of view. Introduce the wind into equation (1.12) by force, namely, add a random value $\xi(t)$:

$$my''(t) = mg - ky(t) + m\xi(t). \tag{1.13}$$

If the vertical component of the wind force $m\xi(t)$ is known, the differential equation (1.13) can be solved. However, $\xi(t)$ is just a symbol which does not solve the problem and hides it in a new symbol. This term $\xi(t)$ will be useful if it is described by the rules used in the probability theory. Below, we outline a simple scheme based on the standard course of probability.

Following the description of the normal distribution we can describe

[9]The cost of electric energy at night is usually lower than during day. Moreover, the cost at night can be negative. Hence, the economic problem to minimize costs complicates the energy problem.
[10]http://eaps4.mit.edu/research/Lorenz/Butterfly_1972.pdf

the random function $\xi(t)$ by two constants, the mathematical expectation $a = E[\xi(t)]$ and the standard deviation $\sigma = \sqrt{E[\xi^2(t)] - E^2[\xi(t)]}$. In the considered case, a expresses the expected averaged wind force acting on the body at time t. Let it be equal to zero that corresponds to the symmetrically expected gusts in the y direction. Application of the expectation operator to (1.13) yields

$$mY''(t) = mg - kY'(t), \tag{1.14}$$

where $Y(t) = E[y(t)]$ denotes the expected trajectory. Next, equation (1.14) can be solved similarly to equation (1.12).

In order to estimate the deviation of the random function $y(t)$, first, we estimate the term $m\xi(t)$ in (1.13) by the bounds $-C\sigma \leq m\xi(t) \leq C\sigma$ where the positive constant C is determined by the chosen confidence interval estimate. Next, we substitute $\pm C\sigma$ in (1.13) instead of $m\xi(t)$ and obtain two ODE

$$mY''_{\pm}(t) = mg - kY'_{\pm}(t) \pm C\sigma. \tag{1.15}$$

Ultimately, we can estimate the expected trajectory and its bounds within the chosen confidence interval as displayed in Fig.1.2.

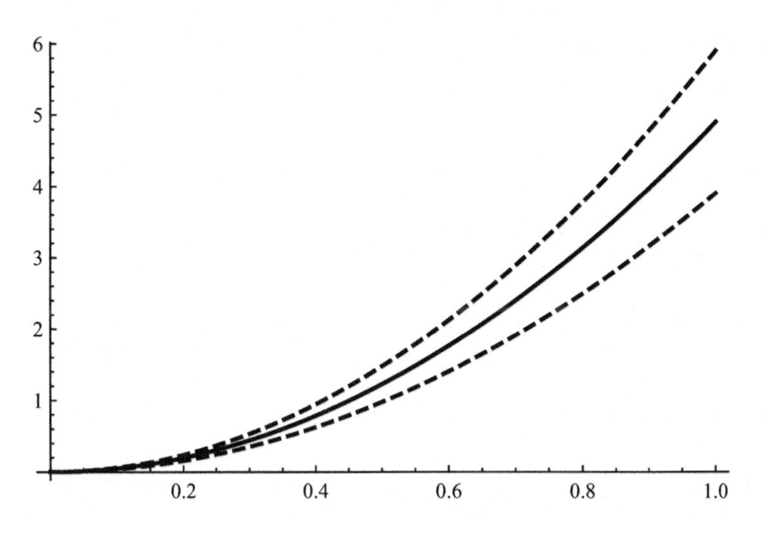

FIGURE 1.2: The expected trajectory (solid line) and its bounds (dashed line).

b) Calculate the mass of the cylinder having various circular sections expressed by the shape function $S = S(x)$ for $x \in [a, b]$ (see Fig.1.3). It is assumed that the cylinder consists of the homogeneous material of the density ρ. Let us divide the cylinder into small cylinders with the heights Δx_i $(i = 1, 2, \cdots, n)$. The mass of the i-th cylinder is equal to $\Delta m_i = \rho\pi S^2(\xi_i)\Delta x_i$, where the point ξ_i is arbitrarily chosen in (x_i, x_{i+1}). If the latter interval is sufficiently small, its mass Δm_i is calculated quite accurately. The total mass

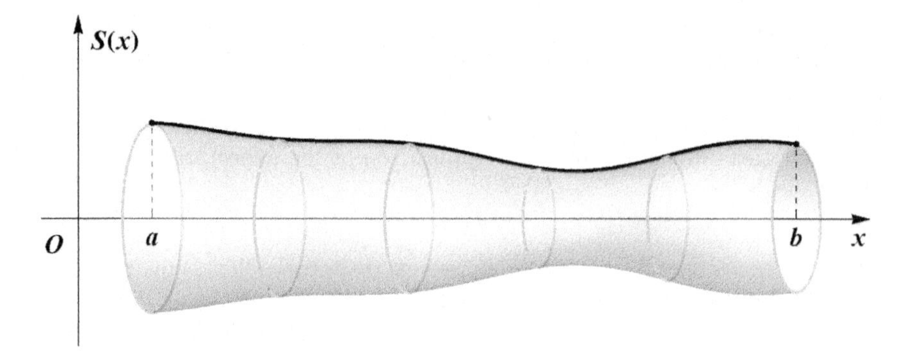

FIGURE 1.3: Radius $S(x)$ of the circular section at the point x.

of the cylinder is approximately equal to

$$M_n = \sum_{i=1}^{n} \Delta m_i = \rho\pi \sum_{i=1}^{n} S^2(\xi_i)\Delta x_i. \tag{1.16}$$

Thus, the mass of the cylinder is calculated by the *discrete model.*

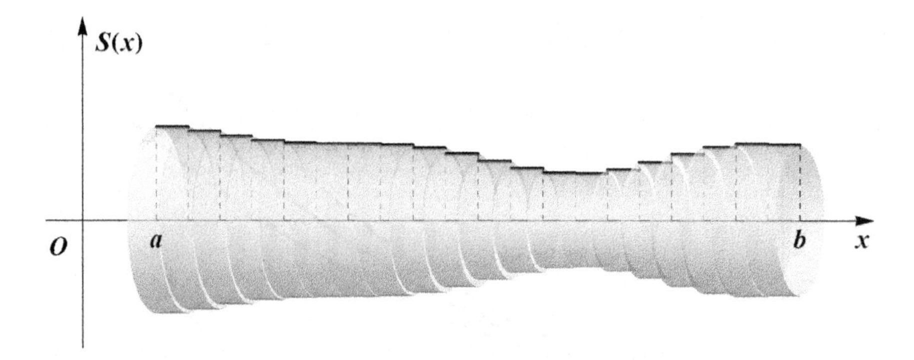

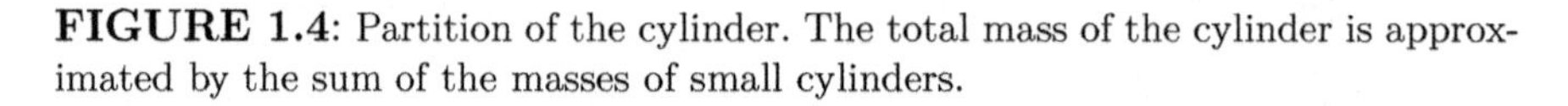

FIGURE 1.4: Partition of the cylinder. The total mass of the cylinder is approximated by the sum of the masses of small cylinders.

Let the value $\delta = \max_i \Delta x_i$ tend to zero, i.e., $[a, b]$ is divided into small intervals whose lengths become smaller and smaller. Then the sum (1.16) tends to a value M, which can be considered as the exact total mass of the cylinder. The sum M_n is called the *Riemann sum* in calculus. The limit M is called the *definite integral* of the function $\rho\pi S^2(x)$ on the interval (a, b):

$$M = \int_a^b \rho\pi S^2(x)\,\mathrm{d}x. \tag{1.17}$$

This integral presents the *continuous model* to calculate the mass.

Application of these two models yields a method to calculate a value (mass) of the complicated objects via small simple ones.

Principle of transition: *continuous* $\leftrightarrow$ *discrete*. *To divide a continuous object into small parts, to apply a simple formula to every part, to calculate the sum for all the parts and get back to the continuous object through the limit operation.*

One can divide an object under the study into parts in space and in time. Partition by the mass $m = \rho x$, where ρ denotes the linear density, is equivalent to partition in space on x.

In mathematics, the principle of transition is the benchmark of the Riemann integral theory including multidimensional, curvilinear and surface integrals. In continuum mechanics this principle is used in the framework of calculus and theory of partial differential equations. Real medium is discrete on the molecular level. Hence, the real mass of the cylinders is rather given by formula (1.16). However, summation over all molecules is practically impossible and integration is more simple than summation in this case. This does not concern nanotechnology when few molecules are taken in consideration.

Remark 1.2. The above example can help to delineate the difference between pure mathematical and practical approaches. Assume that the radius of the body sections presented in Fig.1.3 is described by the Dirichlet function

$$D(x) = \begin{cases} 1, \text{ if } x \text{ is irrational,} \\ \\ 0, \text{ if } x \text{ is rational.} \end{cases} \tag{1.18}$$

There are rational and irrational numbers in every interval of the real axis. If all the points ξ_i are irrational in the sum (1.16), then $D(\xi_i) = 1$ and the total mass of the cylinder is equal to $\rho \sum_{i=1}^{n} \Delta x_i = \rho(b-a)$. On the other hand, if the points ξ_i are rational, then $D(\xi_i) = 0$. Thus, the total mass of the cylinder vanishes. This means that the cylinder is so porous that it does not posses a mass; it is between 0 and $\rho(b-a)$. In this book, we take the side of applied mathematicians. Therefore, we do not consider such strange objects without any mass. However, such examples are very interesting in pure mathematics since they yield new mathematical topics for investigations[11].

c) Linear models are frequently used in practice since they are simple and properly describe real phenomena. The definition of the linear model can be given in terms of the linear operator (function):

Definition 1.1. Operator $L : X \to Y$ is called linear if for any $x, x_1, x_2 \in X$ and an arbitrary real (complex) number λ

i) $L(x_1 + x_2) = L(x_1) + L(x_2)$;

ii) $L(\lambda x) = \lambda L(x)$.

[11]Frequently, an independently developed pure abstract theory has serious applications many years later after its creation.

Example 1.2. The function $y = ax$ is a linear operator $L : \mathbb{R} \to \mathbb{R}$.

Example 1.3. The differentiation operator $\frac{d}{dt} : f \mapsto f'$ is a linear operator. This is true due to the rules $\frac{d}{dt}(f_1 + f_2) = \frac{df_1}{dt} + \frac{df_2}{dt}$ and $\frac{d}{dt}(\lambda f) = \lambda \frac{df}{dt}$.

Example 1.4. If you conduct a uniaxial tensile test on an elastic material you can observe that the force σ per surface of unit area acting on the material is proportional to its deformation $\epsilon = \frac{\Delta \ell}{\ell}$ where $\Delta \ell$ denote displacement of the specimen and ℓ its length. This is Hooke's law which can be written in the form $\sigma = k\epsilon$ where the constant k depends on the material. Linear Hooke's law is valid for small deformation ϵ. Plastic transformations can occur for a large displacement when the coefficient k depends on ϵ. This yields the nonlinear law $\sigma = k(\epsilon)\epsilon$ under simple loading. It is interesting to note that rubber and a metal are elastic materials but with different values of k.

Cybernetics is a multidisciplinary science based on the *black box* when one considers an object which transforms inputs $x \in X$ to outputs $y \in Y$ and is not interested in the interior of the transformation box (Fig.1.5).

FIGURE 1.5: Illustration of the black box. A set of inputs and outputs determines the object without any study of the interior structure.

The rule of the transformation $K : x \to y$ is interesting in cybernetics (roughly speaking, software structure of K, not its hardware). That is why this object is called the black box. Such an approach is frequently used in the theory of signals. A signal $x(t)$ depending on time t is transformed into $y(t)$ by the integral operator:

$$(Kx)(t) := y(t) = \int_0^T k(\tau, t)x(\tau)\, d\tau. \tag{1.19}$$

One can check that the operator (1.19) is linear. Let the kernel have the special structure $k(\tau, t) = k(\tau - t)$. Then, the operator

$$(Kx)(t) := y(t) = \int_0^T k(\tau - t)x(\tau)\, d\tau \tag{1.20}$$

is called *the convolution* of $k(\tau)$ and $x(\tau)$. The transformation $K : x \to y$ can be non-linear [23]

$$(Kx)(t) := y(t) = \int_0^T k(\tau, t, x(\tau))\, d\tau, \tag{1.21}$$

where $k(\tau, t, x)$ is a function of three variables.

1.3 Stability of models

One of the most important characteristic of the model is its stability. Mechanical illustration of stability is displayed in Fig.1.6. Stability can be discussed on a parameter and in time.

FIGURE 1.6: Stable (a) and not stable (b) locations of a ball

Consider the function (1.5) which is continuously dependent on the initial velocity v (parameter of the model) in the closed interval $[0, T]$. This means that for any $t \in [0, T]$ and for sufficiently small changes of v_0 the function (1.5) has small increments. More precisely, let v_1 and v_2 be two such values of the initial velocity that $|v_1 - v_2| < \varepsilon$ for any fixed positive number ε. Let $y_1(t)$ and $y_2(t)$ be functions of the form (1.5) with the velocities v_1 and v_2, respectively. Then

$$\max_{[0,T]} |y_1(t) - y_2(t)| = \max_{[0,T]} |v_1 t - v_2 t| = T|v_1 - v_2| < T\varepsilon.$$

This inequality implies that for sufficiently small $\varepsilon > 0$, the maximal value of the difference $|y_1(t) - y_2(t)|$ for $t \in [0, T]$ is also small, i.e., the discussed problem is stable for v_0.

On the other hand, the time T can be a large number, it can even be equal to $+\infty$. Then $\max_{t\in[0,+\infty]} |y_1(t) - y_2(t)| = +\infty$ for $|v_1 - v_2| < 0.000001$. This means that the problem is unstable in v.

A similar situation takes place for weather prediction (see the footnote on page 16 with the reference to Lorenz's paper and the advanced book [6]). Equations of the air flow on the Earth scale can be stable. Let us say for simplicity that a function $y(t)$ describing the air flow satisfies inequality

$$\max_{[0,T]} |y_1(t) - y_2(t)| < c(T)\varepsilon.$$

Here, the constant $c(T)$ depends on the prediction time T. The value $c(T)$ can be bounded for T which lasts for about a week, what yields well estimations for $y(t)$. However, $c(T)$ abruptly increases for greater T in such a way that $c(T)\varepsilon$ also increases. The stable process becomes unpredictable.

In main, stability is treated as T tends to infinity. Linear homogeneous ODE with constant coefficients have solutions presented as a sum of modes. Each mode has the form $y(t) = P(t)\exp(\lambda t)$ where $P(t)$ is a polynomial, $\lambda = \alpha + i\beta$ is a complex constant and i denotes the imaginary unit. A mode is called stable if it is bounded as $t \to +\infty$. Euler's formula $\exp(\lambda t) = e^{\alpha t}(\cos \beta t + i \sin \beta t)$ yields

Theorem 1.1. *Let $P(t)$ be a non-constant polynomial. A mode $y(t) = P(t)\exp(\lambda t)$ is stable if and only if* $\text{Re}\,\lambda = \alpha < 0$.

Formally, one can divide models into stable and unstable ones. But a physicist will not permit us to do it referring to the principle of stability:

Principle of stability. *A mathematical model must be stable.*

The study of the unstable problem should not give reasonable results since small exterior changes or small changes in the initial data can yield large deviation in the obtained results. But at the beginning of the study no one knows whether the model is stable. Therefore, one should first develop a model, investigate it and only after its verification it is possible to claim its stability. It follows from the above examples that the same model can be stable and unstable. It is worth investigating the conditions under which the problem is stable.

Concluding remarks and further reading concerning this section. Not everyone accepts the principle of stability in such a form and investigates unstable processes following the principle by Dirac stated on page 126. This can be reasonable, for instance, in biological and chemical pattern formations [22, 33]. In the theory of turbulence, the amplitude of the average velocity can be considered when the local velocity is unpredictable [2].

Various mathematical approaches to stability are discussed in courses of ODE and PDE (see Sec.4.3, Sec.4.8 and [5], [41], [33], [22]). The numerical stability is discussed in Sec.8.6.1 ([35], [52]).

1.4 Dimension, units, and scaling

1.4.1 Dimensional analysis

Mathematical models contain measurable values. One cannot say that a dimension value is large or small because it depends on the units in which it is expressed. For example, the length L can be measured in meters (m), centimeters (cm), miles etc. The length L will be called an abstract dimension and the length m a special one. One can treat L as a variable which takes special values m, cm, etc., which can be written as $L = m$. Everybody may introduce his/her own dimension unit. It is only necessary to relate it to the accepted units. The following three dimensions of units are accepted in physics and technology: SI, CGS and English units. Besides the length L we introduce the abstract units of mass M and time T. The velocity has the abstract unit $\frac{L}{T}$ and the special one with $L = m$ and $T = sec$ where sec denotes second. It is convenient to introduce brackets for the units. For instance, $[\rho] = ML^{-3}$ means that the density ρ is considered in dimension units and measured in mass per volume. Similarly, $[5\frac{kg}{m^3}] = kg\ m^{-3}$ means that the density is measured in kilo per spatial meter and is equal to 5 in these units. The expression $[\rho'] = 1$ denotes that ρ' is measured in dimensionless units.

Consider the example from Sec.1.1.1 in the abstract units $[m] = M$, $[t] = T$, $[y] = L$, $v_0 = LT^{-1}$, $[g] = LT^{-2}$. We are interested in the units which appear in equation (1.5). The first term $\frac{gt^2}{2}$ on the right-hand side has the dimension $[\frac{gt^2}{2}] = [g][t]^2 = LT^{-2}T^2 = L$. The second term v_0t on the right-hand side has the dimension $[v_0t] = [v_0][t] = LT^{-1}T = L$. Hence, the right-hand side contains the sum of two values of the same dimension L, that confirms the proper dimensional usage of the expression $\frac{gt^2}{2} + v_0t$. The left-hand side of the equation also has the dimension L. Everything is alright. Checking the dimensions yields a simple way to check an equation. For instance, if one develops a model and gets equations $L = L + M$ or $T = LT^2$, this gives a signal that something is wrong because a meter cannot be compared to a kilo.

Principle of units. *It is forbidden to add camels to tractors. But it is possible to multiply them.*

The units of derivatives can be obtained from the definition of the derivative through the limit of the difference quotients. For instance, $[y] = L$ implies $[y'] = LT^{-1}$ and $[y''] = LT^{-2}$. We now proceed with considering the units of equation (1.12)

$$\begin{aligned}[my''] = [mg] + [kv] \ &\Leftrightarrow\ [m][y''] = [m][g] + [k][v] \\ &\Leftrightarrow\ MLT^{-2} = MLT^{-2} + [k]LT^{-1}.\end{aligned}$$

The principle of units holds if and only if $[k]LT^{-1} = MLT^{-2}$. This implies that the unit of the air resistance, the coefficient k, must be equal to MT^{-1}. The coefficient k can be estimated in experiments. Let $M = kg$ and $T = sec$. In order to determine k one has to present the experimental results in such a way that k will be measured in kilo per second.

Dimensional analysis is applied not only to check a dimensional equation but can suggest a type of such an equation. The main fundamental theorem of dimensional analysis is the Pi theorem [41]. We formulate it in the following reduced form

Theorem 1.2. *Let an equation (E) relate N dimensional quantities. There exists such an equation (e) which contains n dimensionless quantities and*
i) equations (e) and (E) are equivalent;
ii) $n \leq N$.

The full form of the Pi theorem is based on the dimension matrix. We explain this notation on the equation (1.5) written in the form

$$f(y, t, g, v_0) \equiv y - \frac{gt^2}{2} - v_0 t = 0. \tag{1.22}$$

The dimensions are given by $[y] = L$, $[t] = T$, $[v_0] = LT^{-1}$ and $[g] = LT^{-2}$. The dimension matrix A is defined by the powers occurred at these equations. It is convenient to write A in the form of the table

	y	t	v_0	g
L	1	0	1	1
T	0	1	−1	−2

Two dimensionless values can be introduced

$$\pi_1 = t^2 \frac{g}{y}, \quad \pi_2 = t\frac{v_0}{y}. \tag{1.23}$$

The value g and v_0 are expressed through other values by equations

$$g = \pi_1 \frac{y}{t^2}, \quad v_0 = \pi_2 \frac{y}{t}. \tag{1.24}$$

Substitution of (1.24) into (1.22) yields

$$1 - \frac{\pi_1}{2} - \pi_2 = 0. \tag{1.25}$$

An experimental verification of (1.22) can be done by measurement of y, t, v_0 and g in arbitrary units. Further, π_1 and π_2 are calculated with (1.23). At the end, the value $1 - \frac{\pi_1}{2} - \pi_2$ has to be compared with zero, i.e., equation (1.25) has to be checked.

1.4.2 Scaling

Each dimensional parameter has its domain of values. Estimations of this domain are useful in the introduction of the dimensionless units by scaling.

The parameters of equation (1.12) can be made dimensionless in the following way

$$y = \ell\tilde{y},\ t = \tau\tilde{t},\ m = m_0\tilde{m}, \tag{1.26}$$

where ℓ, τ, m_0 are characteristic values of the length, time and mass, respectively. For instance, if a book falls from a table, one can assume $\ell = 1cm$, $\tau = 1sek$, $m_0 = 0.1kg$. If a parachutist falls from an plane, one can take $\ell = 700m$, $\tau = 1min$, $m_0 = 10kg$. The choice of the scale can be arbitrary. However, it is better to take reasonable values in order to work with "usual" numbers, to avoid divisions into small numbers following the principle of the balanced computations.

The derivatives y' and y'' after the substitutions (1.26) are transformed as follows

$$\frac{dy}{dt} = \frac{\ell}{\tau}\frac{d\tilde{y}}{d\tilde{t}},\quad \frac{d^2y}{dt^2} = \frac{\ell}{\tau^2}\frac{d^2\tilde{y}}{d\tilde{t}^2}. \tag{1.27}$$

Then equation (1.12) becomes

$$\tilde{y}'' + \tilde{k}\tilde{y}' = \tilde{g}, \tag{1.28}$$

where $\tilde{k} = \frac{k\tau}{m}$ denotes a dimensionless coefficient. This can be proved in the following way

$$[\tilde{k}] = [k]TM^{-1} = MT^{-1}TM^{-1} = 1.$$

Concluding remarks and further reading concerning this section. A short introduction to the unit theory and scaling is presented above. Further extensions and applications can be found in [41].

Exercises

1. Describe the trajectory of the falling object thrown in a horizontal direction.

2. Describe the trajectory of the falling object thrown in a horizontal direction when the land is determined by a surface, for instance by a linear function $y = a_1x_1 + a_2x_2 + b$.

3. Fig.1.1 presents a film frame of the falling object. Prepare an animation of the whole process. Calculate online the velocity and the acceleration of the point. Take into account the air resistance.

   Useful *Mathematica* operators: **`Animate`**, **`Manipulate`**, **`Slider`**, **`TrackedSymbols`**, **`DynamicModule`**. For creating animation in MATLAB refer to Example Box 8.1 on page 175.

As an advanced exercise, you can create slider manipulator in MATLAB. Refer to the documentation and investigate the `uicontrol` operator.

4. Calculate numerically the sums $\sum_{n=1}^{\infty} \frac{1}{n^m}$ for $m = 1, 2, 3, 4, \ldots$.

   Think about the outputs **`Zeta[3]`** and **`Zeta[3.]`** in *Mathematica* for $m = 3$.

5. Let hot coffees and cold cream in Example 1.1 be replaced by cold water with ice and juice at room temperature, respectively. Who drinks colder water?

6. Calculate the mass of the circular cone of height h and radius r with constant density ρ.

7. Are the following operators linear?

   A linear function $y = ax + b$ from $\mathbb{R}$ on $\mathbb{R}$;

   function $Y = AX$ from $\mathbb{R}^n$ in $\mathbb{R}^n$, where A is a $n \times n$ matrix;

   integral operator $\int_0^{\infty} k(\tau, t)x(t)\, d\tau$.

8. Present a graphical interpretation of the stability of the process (1.5) in time when the initial velocity v_0 is measured with an error of 5%.

Chapter 2

Numerical and symbolic computations

2.1	Numerical and symbolic computations of derivatives and integrals	27
2.2	Iterative methods	29
2.3	Newton's method	30
2.4	Method of successive approximations	32
2.5	Banach Fixed Point Theorem	34
2.6	Why is it difficult to numerically solve some equations?	37
	Exercises	39

2.1 Numerical and symbolic computations of derivatives and integrals

The derivative and *integral* are the most important notations of the mathematical analysis. The required theoretical background can be found in standard calculus textbooks cited in the introduction.

Mathematica as well as other packages easily operate with derivatives:

```
In[1]:= f[x_] := e^x Cos[x]^2

In[2]:= D[f[x], x]

Out[2]= e^x Cos[x]^2 - 2 e^x Cos[x] Sin[x]
```

The definite integral is introduced as the area enclosed between the graph of the continuous positive function and the x-axis making use of the Riemann sum and the principle of transition *continuous* $\leftrightarrow$ *discrete*. In Sec.1.2, the definite integral (1.17) is introduced as the mass of the cylinder in the same way. Calculate in *Mathematica* the integral from the above introduced function f on the interval $[0, \frac{\pi}{2}]$:

```
In[3]:= Integrate[f[x], {x, 0, π/2}]

Out[3]= 1/5 (-3 + 2 e^(π/2))
```

We recommend the cited textbooks devoted to approximate computations of derivatives and integrals (see for instance [35]). It is worth studying Help of *Mathematica* to try the operator **`Derivative`** and **`NIntegrate`**.

MATLAB Example Box 2.1

Let us investigate `diff` and `integral` functions of MATLAB. The following scripts compare a symbolic derivative with the numerical one.

```
function script04(h)
    % if h is not provided set the default value
    if nargin==0
        h = 0.01;   % default step size
    end

    % define function f
    f = @(x) exp(x) .* cos(x).^2;

    % define domain sample points and compute values of f
    X = 0:h:pi/2;
    Y = f(X);

    % compute approximate derivative
    Y_diff = diff(Y)/h;

    % compare analytical and numerical solutions
    f_prime = @(x) exp(x).*(cos(x).^2 -2*cos(x).*sin(x));
    Y = f_prime(X);
    plot(X(:,1:length(Y_diff)), Y_diff, '-k', X, Y, 'r');
    grid on
```

One can observe the dependence of accuracy on `h`.

```
>> script04(0.1)
```

step size h = 0.01

step size h = 0.10

As another example, let us integrate the function f numerically

```
>> f = @(x) exp(x) .* cos(x).^2;
>> f_integr = integral(f, 0, pi/2)
f_integr =
    1.3242
```

and compare it with the numerical value of Out[3] in *Mathematica* above:

```
>> 1/5*(-3 + 2*exp(pi/2))
ans =
    1.3242
```

2.2 Iterative methods

Iteration is a rule which can be successively applied infinitely many times. It produces a sequence of objects (numbers) that follow a given pattern.

For instance, the Fibonacci numbers are the sequence $\{F_n\}$ $(n = 1, 2, \cdots)$ defined by the iterations

$$F_n = F_{n-1} + F_{n-2}, \quad n = 3, 4, \ldots, \tag{2.1}$$

i.e., the next number F_n is the sum of the two previous ones F_{n-1} and F_{n-2}. The first two numbers are given as $F_1 = F_2 = 1$. Here, the rule (2.1) expresses the iteration and the iterative method. The Fibonacci numbers are $1, 1, 2, 3, 5, 8, 13, 21, 34, 55, \ldots$.

In general the iterative method is defined through an operator A which describes the rule how to construct an element (number) x_n by means of the previous ones $x_1, x_2, \ldots, x_{n-1}$

$$A : (x_1, x_2, \ldots, x_{n-1}) \mapsto x_n. \tag{2.2}$$

The iterative method is called *convergent* if the sequence $\{x_n\}$ converges as $n \to \infty$. Frequently, the operator A explicitly determines x_n only by x_{n-1}

$$x_n = Ax_{n-1}, \quad n = 2, 3, \ldots. \tag{2.3}$$

Then, the first element x_1 has to by given. The iterative method (2.3) is called *direct*. The *indirect* iterative method is formally determined by the expression

$$Bx_n = Cx_{n-1}, \quad n = 2, 3, \ldots, \tag{2.4}$$

where B and C are given operators. The rule (2.3) directly determines x_n through x_{n-1}. In order to determine x_n by (2.4) we have additionally to solve the equation (2.4) on x_n. If such an equation has a unique solution, i.e., the inverse operator B^{-1} is correctly defined, the indirect method can be written as the direct method, $x_n = B^{-1}Cx_{n-1}$. It is worth noting that in practice frequently B^{-1} is just a symbol helping us to understand the general picture. The art of iterative methods consists in the decomposition $A = B^{-1}C$ where it is easy to solve the equation $Bx = b$ on x and to get the convergent method (2.4). Below, we discuss the main computational schemes based on the iterations.

2.3 Newton's method

Let a function $f : (a, b) \to \mathbb{R}$ be continuously differentiable on the open interval (a, b). Newton's method can be applied to the following equation

$$f(x) = 0, \tag{2.5}$$

where x is unknown. *Mathematica* contains operators to solve such an equation. The following operator numerically finds a root with zero approximation $x_0 = 5$.

```
In[1]:= FindRoot[x^3 e^x + Cos[x], {x, 5}]

Out[1]= {x → -5.47198}
```

We proceed with a discussion of the theoretical background of Newton's method since it has applications not only in equations in one variable but in general operator equation where f can be an integral operator of the type (1.19).

It follows from the definition of the derivative of f that

$$f(x_0 + h) = f(x_0) + hf'(x_0) + o(h), \quad h \to 0, \tag{2.6}$$

where $o(h)$ denotes a value infinitesimally small compared to h, i.e., $\lim_{h\to 0} \frac{o(h)}{h} = 0$. Let x_0 be a given number (zero approximation) and $f'(x_0) \neq 0$. Let h be an unknown correction to the solution $x = x_0 + h$. Then

$$f(x_0) + hf'(x_0) \approx 0 \quad \Leftrightarrow \quad h \approx -\frac{f(x_0)}{f'(x_0)} \quad \Leftrightarrow \quad x \approx x_0 - \frac{f(x_0)}{f'(x_0)}.$$

The latter relation can be considered as the first order approximation

$$x_1 = x_0 - \frac{f(x_0)}{f'(x_0)}. \tag{2.7}$$

The k-th approximation has the form

$$x_k = x_{k-1} - \frac{f(x_{k-1})}{f'(x_{k-1})}, \quad k = 1, 2, \ldots \tag{2.8}$$

An iterative process to find higher order approximations has been constructed. The main problem consists of its convergence. If the limit $x = \lim_{k\to\infty} x_k$ exists, calculation of the limit in (2.8) as $k \to \infty$ yields (2.5).

It follows from the geometric observations that Newton's method can be convergent and can be divergent, that depends on the choice of zero approximation x_0 (see Fig.2.1 and Fig.2.2).

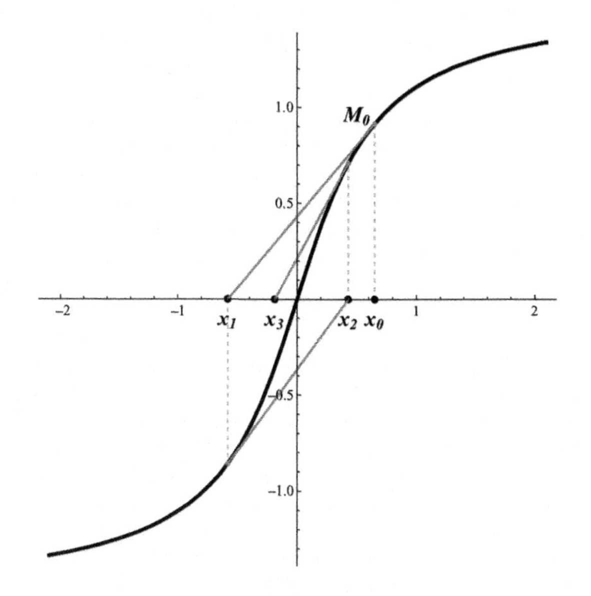

FIGURE 2.1: Take a zero approximation x_0 and construct the point $M_0(x_0, f(x_0))$. The tangent line at this point to the graph of a function $y = f(x)$ passes through the axis OX in the point x_1 considered as the first approximation. In accordance with the geometric interpretation of the derivative we have $f'(x_0) = \tan\alpha$. On the other hand, we have $\tan\alpha = \frac{f(x_0)}{x_1 - x_0}$ by consideration of the triangle $M_0x_0x_1$. Hence, $f'(x_0) = \frac{f(x_0)}{x_1 - x_0}$ that yields equation (2.7). A point x_2 is constructed in the same way through x_1 and so on. In the present example, the iterations converge to a root of equation (2.5).

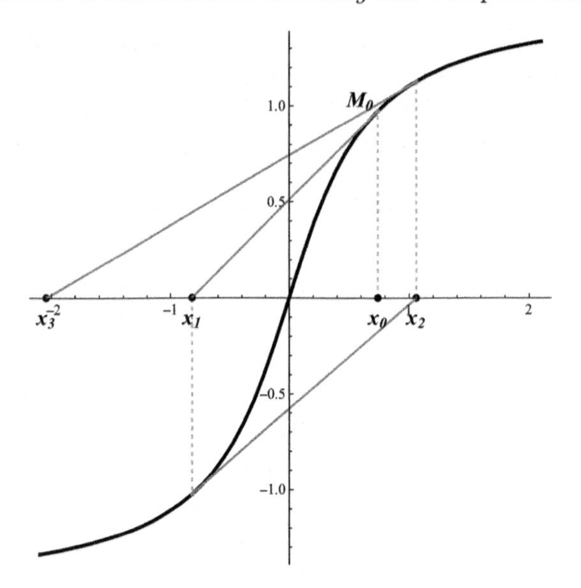

FIGURE 2.2: Successive approximations $x_1, x_2, \ldots$ are constructed in accordance with the rule described in Fig.2.1. In this example, the sequence $\{x_k\}$ diverges.

MATLAB Example Box 2.2

In MATLAB one can use `fzero` function as follows

```
>> f = @(x) x^3 * exp(x) + cos(x);
>> x0 = 5;      % initial point
>> fzero(f, x0)
ans =
    -1.1064
>> fzero(f, -4)
ans =
    -5.4720
```

2.4 Method of successive approximations

Consider equation (2.5) written in the form

$$f(x) + x = x \quad \Leftrightarrow \quad x = \varphi(x), \tag{2.9}$$

where the new function $\varphi(x) := f(x) + x$ is introduced.

In order to determine reasons why iterations can be convergent or divergent we consider a simple example of equation (2.9):

$$x = \sqrt[3]{x^2} + 2, \tag{2.10}$$

This equation has a unique solution. Let us try to approach this solution by successive approximations

$$x_k = \sqrt[3]{x_{k-1}^2} + 2, \quad k = 1, 2, \ldots \tag{2.11}$$

Use *Mathematica*

```
In[1]:= x_k_ := ∛(x_(k-1)^2) + 2
```

```
In[2]:= x0 = 5.; Table[x_k, {k, 0, 24, 2}]
        x0 = -100.; Table[x_k, {k, 0, 24, 2}]

Out[2]= {5., 4.89432, 4.87809, 4.87559, 4.8752, 4.87514, 4.87513,
         4.87513, 4.87513, 4.87513, 4.87513, 4.87513, 4.87513}

Out[3]= {-100., 10.2147, 5.55675, 4.97728, 4.89084, 4.87756, 4.8755,
         4.87519, 4.87514, 4.87513, 4.87513, 4.87513, 4.87513}
```

One can see that x_k tends to 4.87513 as $k \to \infty$.

Now, solve the same equation by means of another method. Equation (2.10) is equivalent to equation

$$x = \sqrt{(x-2)^3}. \tag{2.12}$$

Let us try to solve it by successive approximations using *Mathematica*

$$x_k = \sqrt{(x_{k-1} - 2)^3}, \quad k = 1, 2, \cdots . \tag{2.13}$$

We have

```
In[4]:= x_k_ := √((x_(k-1) - 2)^3)
```

```
In[5]:= x0 = 6.; Table[x_k, {k, 0, 14, 2}]
        x0 = 150.; Table[x_k, {k, 0, 14, 2}]

Out[5]= {6., 14.6969, 284.36, 326613., 2.55018×10^12,
         8.21834×10^27, 6.43079×10^62, 2.08255×10^141}

Out[6]= {150., 76271.9, 9.66708×10^10,
         5.21092×10^24, 4.10258×10^55, 1.34703×10^125,
         3.476166324816573×10^281, 2.934108929590624×10^633}
```

The sequence x_k diverges in this case.

MATLAB Example Box 2.3

One can compute successive approximations using `for` loop in MATLAB as well.

```
function script06()
    x = 150; % initial point
    for k = 2:21
        % compute the next approximation
        x = x^(2/3) + 2;
        fprintf('k = %-3d -> %f\n', k, x);
    end
```

The following output will be printed.

```
k = 2   -> 30.231081
k = 3   -> 11.704410
k = 4   -> 7.155051
k = 5   -> 5.713144
k = 6   -> 5.195829
k = 7   -> 4.999875
k = 8   -> 4.923969
k = 9   -> 4.894300
k = 10  -> 4.882662
            . . .
k = 19  -> 4.875131
k = 20  -> 4.875130
k = 21  -> 4.875130
```

2.5 Banach Fixed Point Theorem

Definition 2.1. Let a constant α satisfy the inequality $0 < \alpha < 1$. Let (X, ρ) denote a metric space with the metric ρ. An operator $A : X \to X$ is called a contraction operator if $\rho(Ax, Ay) \le \alpha\rho(x, y)$ for any $x, y \in X$.

The above definition can be simplified by consideration of the operator on the real axis $\mathbb{R}$. So, one can skip Definition 2.1 and pay attention to the following one.

Definition 2.2. Let a constant α satisfy inequality $0 < \alpha < 1$ and (a, b) denote an open interval. A function $\varphi : (a, b) \to (a, b)$ defined in the metric space $(\mathbb{R}, |\cdot|)$ with the metric $|x - y|$ is called a contraction if $|\varphi(x) - \varphi(y)| \le \alpha|x - y|$ for any $x, y \in \mathbb{R}$.

Theorem 2.1 (Banach Fixed Point Theorem). *Let A be a contraction operator in the Banach (complete metric) space (X, ρ). Then equation*

$$x = Ax \tag{2.14}$$

has one and only one solution which can be found by successive approximations

$$x_k = Ax_{k-1}, \quad k = 1, 2, \ldots \tag{2.15}$$

convergent in X for any x_0.

Proof for equation $x = \varphi(x)$ follows from estimations for the recurrent sequence $x_k = \varphi(x_{k-1})$. We have

$$|x_{k+1} - x_k| \leq \alpha|x_k - x_{k-1}| \leq \alpha^2|x_{k-1} - x_{k-2}| \leq \cdots \leq \alpha^k|x_1 - x_0|.$$

Consequently, for any fixed $m \geq 1$

$$\begin{aligned} |x_{k+m} - x_k| &\leq |x_{k+m} - x_{k+m-1}| + |x_{k+m-1} - x_{k+m-2}| + \cdots + |x_{k+1} - x_k| \\ &\leq (\alpha^{k+m-1} + \cdots + \alpha^k)|x_1 - x_0| \leq \frac{\alpha^k}{1-\alpha}|x_1 - x_0|. \end{aligned}$$

This implies that Cauchy's criterion (recall calculus) holds for the sequence $\{x_k\}$. There exists the limit $x = \lim_{k\to\infty} x_k$ since the space X is complete (roughly speaking, X does not have holes). It follows from the Definition 2.2 that f is continuous. This yields $\lim_{k\to\infty} x_{k+1} = \varphi(\lim_{k\to\infty} x_k) \Longleftrightarrow x = \varphi(x)$.

Let x and y be two solutions of equation $x = \varphi(x)$. Then

$$|x - y| = |\varphi(x) - \varphi(y)| \leq \alpha|x - y|, \text{ hence } (1-\alpha)|x - y| \leq 0 \Rightarrow \quad x = y$$

that yields uniqueness of solution.

The theorem in $\mathbb{R}$ is proved.

Corollary 2.1. *It follows from the Banach Theorem that equation* (2.9) *is an equation with a contraction operator if $|\varphi'(x)| < 1$ for any $x \in (a, b)$.*

The operator from (2.10) is a contraction whereas the operator from (2.12) is not a contraction. Simple examples of equations from Sec.2.3 demonstrate different cases that can occur in applications. The example below shows wide possibilities of the method of successive approximations.

Definition 2.3. Let $0 < \lambda \leq 1$. A function $g(x)$ satisfies the Hölder condition on X if

$$|g(x_1) - g(x_2)| \leq C|x_1 - x_2|^\lambda, \quad x_1, x_2 \in X, \tag{2.16}$$

where C is a positive constant not depending on x_1 and x_2.

Example 2.1. Let a function $f(t, x)$ satisfy the Hölder condition in the variable x and α be a number parameter. Consider Cauchy's problem

$$x' = \alpha f(t, x), \quad x(0) = x_0, \tag{2.17}$$

where $x = x(t)$ is an unknown function defined in the interval $[0, T]$ and x_0

is a given number. Application of the integral operator $\int_0^t \cdot\, dt$ to (2.17) yields the integral equation

$$x(t) = \alpha \int_0^t f(\tau, x(\tau))d\tau + x_0, \quad t \in [0, T], \tag{2.18}$$

shortly written in the operator form

$$x = \alpha Ax + x_0. \tag{2.19}$$

Application of the Banach Fixed Point Theorem (Th.2.1) to (2.19) gives $x(t)$ in the series form

$$x = \sum_{k=0}^{\infty} \alpha^k A^k x_0. \tag{2.20}$$

The operator αAx is a contraction in the corresponding functional space [41]. Hence, the series (2.20) converges. The main difficulty in applications of the series (2.20) is iterated integrals encrypted in A^k. However, for small $|\alpha|$ or T a few of terms can give a sufficiently precise solution. For instance, the first order approximation in α has the form

$$x(t) \approx x_0 + \alpha \int_0^t f(\tau, x_0)d\tau, \quad 0 \le t \le T. \tag{2.21}$$

This approximate formula is efficient in evolution problems when T is considered as the step of iterations in time.

Example 2.2 (numerical). Consider the numerical example with $f(t, x) = [1 + (x + t)]^{-1}$ and $x_0 = 1$ in the problem (2.17). The first order approximation is obtained by symbolic computations

```
In[1]:= f[t_, x_] = 1/(1 + (x + t)^2);

In[2]:= X[0] = 1;

In[3]:= X[1][t_] =
          Series[X[0] + α Simplify[∫_0^t f[τ, X[0]] dτ, Assumptions → t > 0],
           {α, 0, 1}] // Normal

Out[3]= 1 + α (-π/4 + ArcTan[1 + t])

In[4]:= f1[t_] = Series[f[t, X[1][t]], {α, 0, 1}] // Normal

Out[4]= 1/(1 + (1 + t)^2) - (2 (1 + t) α (-π/4 + ArcTan[1 + t]))/(1 + (1 + t)^2)^2
```

One can go ahead and calculate the next approximations. For instance, the second order approximation has the form

```
In[5]:= X[2][t_] = X[0] + α Simplify[∫_0^t f1[τ] dτ, Assumptions → t > 0 && α > 0]

Out[5]= 1 + 1/(8 (2 + t (2 + t))) α (-2 π (2 + t (2 + t)) +
          t (2 t + π (2 + t)) α + 4 (4 - t (2 + t) (-2 + α)) ArcTan[1 + t])
```

```
In[6]:= X[2][t_] = Series[X[2][t], {α, 0, 2}] // Normal

Out[6]= 1 + 1/4 α (-π + 4 ArcTan[1 + t]) +
        (α^2 (2 π t + 2 t^2 + π t^2 - 8 t ArcTan[1 + t] - 4 t^2 ArcTan[1 + t]))/(8 (2 + t (2 + t)))
```

Such a scheme of symbolic computations is useful when integrals are calculated in terms of elementary functions.

2.6 Why is it difficult to numerically solve some equations?

In this section, we consider numerically hardly solvable equations by conventional methods. From a purely mathematical point of view these equations are properly stated.

Example 2.3. Consider equation

$$P_n(x) \equiv (x-1)^n(\sin x + 2) = 0 \tag{2.22}$$

with $n = 50$. It is evident that this equation has the unique solution $x = 1$. However, if we try to solve (2.22) numerically, we get the following strange result

```
In[1]:= P[x_, n_] := (x - 1)^n (Sin[x] + 2);

In[2]:= FindRoot[P[x, 50] == 0, {x, 0.}]

Out[2]= {x → 0.868256}

In[3]:= FindRoot[P[x, 50] == 0, {x, 1.03}]

Out[3]= {x → 1.00398}
```

with the warning message "FindRoot: Failed to converge to the requested accuracy or precision within 100 iterations." It is important to understand why we get such an output in order to properly operate with other bad equations. We will use the *principle of microscope* as follows. First, we calculate $P_{50}(0.868256) = 2.67889\ 10^{-44}$ and $P_{50}(1.003978978) = 2.69789\ 10^{-240}$. Numerically, it is zero. Look at the graph of $P_{50}(x)$ in Fig.2.3 near the point $x = 1$. One can see that numerically it vanishes in accordance with the prescribed precision. If the precision holds 10^{-9}, the whole interval $(0.25, 1.75)$ fits for solution to (2.22). If the precision is increased to 10^{-80}, the interval $(0.98, 1.02)$ is taken and so forth.

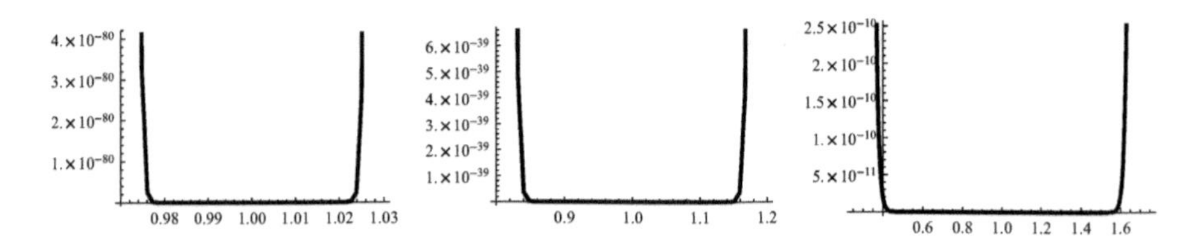

FIGURE 2.3: Graph of $P_{50}(x)$.

MATLAB Example Box 2.4

Let us investigate the same example in MATLAB.

```
>> P = @(x, n) (x-1)^n * (sin(x) + 2);
>> P50 = @(x) P(x, 50);
>> fzero(P50, 1.01)
```

As we can see, the warning is raised and MATLAB suggests a different initial value.

```
Exiting fzero: aborting search for an interval containing a
    sign change
    because NaN or Inf function value encountered during
        search.
(Function value at -1.91711e+06 is Inf.)
Check function or try again with a different starting value.

ans =
   NaN
```

Example 2.4. Consider Kepler's equation having applications in classical celestial mechanics

$$5\cos x - x = 0. \tag{2.23}$$

Let us apply the operator **`FindRoot[5Cos[x]-x==0,{x,6}]`** with the zeroth approximation $x_0 = 6$. *Mathematica* answers "The line search decreased the step size to within tolerance specified by AccuracyGoal and PrecisionGoal but was unable to find a sufficient decrease in the merit function. You may need more than MachinePrecision digits of working precision to meet these tolerances" and suggests $x = 6.08183$. We get the same answer for other zeroth approximation $x_0 = 0$. If we put $x_0 = 4$, we get $x = 6.11574$. The function $y(x) = 5\cos x - x$ is displayed in Fig.2.4 below clarifies the situation. One

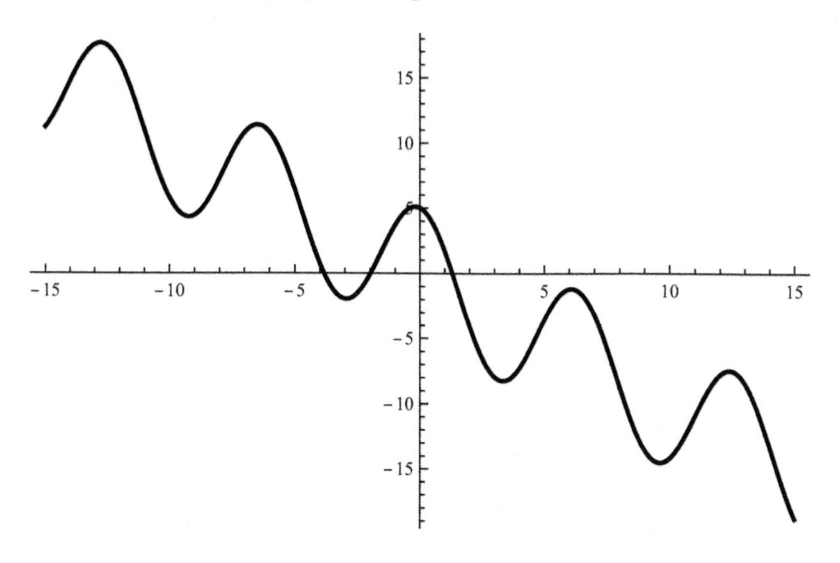

FIGURE 2.4: Graph of $y(x) = 5\cos x - x$

can see that the function $y(x)$ has three roots about -4, -2 and 1. If we put these values as the zeroth approximations, we get the proper roots -3.98583, -1.89152 and 1.34475 without any warning. The suspicious point $x = 6.11574$ arises because the value $y(6.11574)$ is closed to the axis and locally the graph is concave near that point. *Mathematica* tries to improve the result at each step going up, following the principle of stupid computer stated on page 10.

Concluding remarks and further reading. Numerical computations is a separate great topic of applied mathematics arisen in ancient times and extensively developed from the second generation of computers in 1947. Symbolic computations became popular at the end of 20th century with development of the Computer Algebra Systems (CAS). Competition in this field led to the excellent user-friendly packages *Mathematica* used in this book and *Maple*. Theory and applications of numerical computations can be found in many introductory, e.g. [19, 25, 49, 23], and advanced books, e.g. [35, 1]. We refer to the works [14, 17, 18, 50, 59] concerning symbolic computations theory. For an introductory level content and a reference see the official *getting started* online materials of both *Mathematica* and MATLAB, as well as the excellent documentation attached to the desktop applications. There is a rich collection of books on those packages available, e.g. [60, 56, 29, 28, 30, 43, 27].

Exercises

1. Apply the *Mathematica* operators **Derivative**, **Integrate**, and **NIntegrate** in various forms to the functions $\sin mx$, $\exp(-|x|)$, $\exp(-x^2)$, $\frac{\sin(\alpha x)}{x}$. Calculate the corresponding definite integrals $\int_a^b f(x)dx$ for a and b in symbolic form, for the values:

   $a = -1$ and $b = 1$;

   $a = 0$ and $b = +\infty$;

   $a = -\infty$ and $b = +\infty$.

   Explain the obtained results.

   For infinite endpoints in numerical integration with `integral` operator in MATLAB use the value `Inf`.

2. Prepare a computer code to calculate the Fibonacci numbers (2.1).

   To form the sequence $\phi_n = \frac{F_n}{F_{n-1}}$ and compute its limit $\phi = \lim_{n\to\infty} \frac{F_n}{F_{n-1}}$. To check that $\phi = \frac{1+\sqrt{5}}{2}$ is the golden ratio.

3. Apply the *Mathematica* operator **Solve** to equations

   $ax^2 + bx + c = 0,$

   $x^3 + x + 1 = 0,$

   $x^4 + x + 1 = 0,$

   $x^5 + x + 1 = 0,$

   $x^5 + x^2 + 1 = 0.$

   Add the numeric operator **//N** to the obtained results or apply **NSolve**.

4. Apply the operator **FindRoot** to the above equations with various zero approximations including complex numbers. Try to find solutions of non-symbolic equations with `fzero` from MATLAB.

5. Apply **FindRoot** to equations

   $\sin x + x + 1 = 0,$

   $\log x + x + 1 = 0,$

   $0.2 \log x - x + 1 = 0,$

   $10^x x^5 + x^2 + 1 = 10,$

   $(\sin x)^x + x^2 + 1 = 10.$

   Solve the same equations with MATLAB. Check the obtained result by substitution and by graph.

6. Find the maximal root of equations

   $(x-2)\sin\frac{1}{x} = 1,$

   $\sin\frac{1}{x} = \frac{1}{30},$

   $x^4\log\frac{1}{x} = -330000.$

   Hint: If you cannot guess an appropriate zero approximation for an equation $f(x) = 0$, use the principle of microscope. First, construct the graph $y = f(x)$ using the operator **Plot**. Calculate $\lim_{x\to+\infty} f(x)$ to investigate the behavior of $y = f(x)$ at infinity, perhaps with the operator **Limit**. Select such an interval on the x-axis that zeros of $y = f(x)$ become visible. Reduce the interval if it is necessary to take an appropriate zero approximation.

   Apply an analogous scheme using MATLAB with `plot` and `fzero` operators by observing the plot on finite intervals.

7. Using **FindRoot** in *Mathematica* solve equation

   $\left(\log(x-1000) + \log\frac{1}{x-1000}\right)\exp(x^2+100x+546) = 0.$

   Hint: Apply the operator **Simplify** with **Assumption** "$x > 1000$".

8. Introduce the function $f(x,d,M) = \sum_{m=-M}^{M}\exp[-\frac{(x-m)^2}{d}]$. Solve equations

   $f(0.1x, 3, 10) = 3.$

   Find positive roots of equation

   $xf(x, 2, 100) + 0.01 = 0.$

9. Introduce the function $g(x,d,M) = \sum_{m=-M}^{M}\exp[-\frac{(x-m)^2}{d}]\sin mx$. Find all the roots of equation $g(x, 5, 10) = 3$ on the interval $(0, 10)$.

10. Solve the differential equation $y'(t) = y^2(t) + \sin(t)$ with the initial condition $y(0) = 0$ by the method of successive approximations.

11. Solve the integral equation

$$y(t) = \int_0^t t\sin[y(t)]\,dt + t, \quad 0 \le t \le T$$

   by the method of successive approximations for various numerical values of T.

Part II

Basic Applications

Chapter 3

Application of calculus to classic mechanics

3.1 Mechanical meaning of the derivative 45
3.2 Interpolation 46
3.3 Integrals 52
3.4 Potential energy 54
Exercises 55

3.1 Mechanical meaning of the derivative

Derivatives are studied in the standard course of calculus. We are now interested in applications of derivatives in classical mechanics. Consider a one-dimensional (1D) path of the material point on the real axis. It can be described by a function of time $S = S(t)$ where $t \in [0, T]$. Let a point t_0 lie in the interval $(0, T)$, Δt denotes the increment of time and $t_0 + \Delta t$ belongs to $(0, T)$. Then $\Delta S = S(t_0 + \Delta t) - S(t_0)$ expresses the increment of the path in time Δt beginning from t_0. The average velocity of the material point over Δt is given by the ratio $\frac{\Delta S}{\Delta t}$. Let the ratio $\frac{\Delta S}{\Delta t}$ tend to a value $v(t_0)$ as $\Delta t \to 0$. The value $v(t_0)$ is called the instantaneous velocity.

Mechanical meaning of the derivative. *A derivative of a path (trajectory) $S(t)$ with respect to time expresses the instantaneous velocity $v(t)$:*

$$v(t) = S'(t) = \frac{\mathrm{d}S}{\mathrm{d}t}.$$

A derivative of a velocity $v(t)$ expresses an acceleration $a(t)$, hence the second derivative of $S(t)$ gives $a(t)$:

$$a(t) = v'(t) = S''(t) = \frac{\mathrm{d}^2 S}{\mathrm{d}t^2}.$$

The principle of transition *continuous* $\leftrightarrow$ *discrete* for a finite interval of time is used in the interpretation of the derivative. First, the continuous path $S(t)$ is considered. Next, the finite ratio $\frac{\Delta S}{\Delta t}$ is constructed as a discrete object. Further, the derivative $v(t) = S'(t)$ (continuous object) is introduced as a limit of a discrete object as $\Delta t \to 0$.

The principle of transition *continuous* $\leftrightarrow$ *discrete* for derivatives is based on the replacement of the rate of change of the function with the derivative and vice versa. The argument can be other than time. For instance, the mass of the cylinder on the interval Δx is equal to Δm. Then $\frac{\Delta m}{\Delta x}$ gives the average density on Δx. Therefore, the derivative $\frac{dm}{dx} \approx \frac{\Delta m}{\Delta x}$ gives the point density.

3.2 Interpolation

Let a function be given in the form of a table, i.e., in the form of a set (x_i, y_i) $(i = 1, 2, \ldots, n)$. Is it possible, and how does one calculate a derivative of such a function? Everything depends on the type of data. If the data represents exact values of a function, it is better to use an interpolation. *Interpolation* is such an analytical reconstruction of the function f that the values of this function coincide with the given data, i.e., $y_i = f(x_i)$ $(i = 1, 2, \ldots, n)$. Usually polynomial interpolations are used. Sets of polynomials are also used (spline interpolation approximation). The most popular interpolation is *Lagrange's interpolation* or *polynomial interpolation* (introduced by Waring in 1779) by polynomials of degree at most $n - 1$:

$$P(x) = \sum_{i=1}^{n} \left(y_i \prod_{k \neq i} \frac{x - x_k}{x_i - x_k} \right), \tag{3.1}$$

where $\prod_{k \neq i}$ denotes the product over $k = 1, 2, \ldots, n$ except $k = i$. In order to use formula (3.1) we need $O(n^2)$ operations[1] (addition and multiplication). For computations, the same formula can be written in a more effective form [12]

$$P(x) = \frac{\sum_{j=1}^{n} \frac{w_j y_j}{x - x_j}}{\sum_{i=1}^{n} \frac{w_j}{x - x_j}}, \tag{3.2}$$

where

$$w_j = \left[\prod_{k \neq j} (x_j - x_k) \right]^{-1}. \tag{3.3}$$

First, the division in (3.2) is balanced, i.e., the values of the numerator and denominator have the same order. Second, the number of operations to calculate (3.2) has the order $O(n)$.

---

[1] Here, $O(N)$ means the value of order N as N tends to infinity. For instance, N and $2N$ have the same order $O(N)$, but $0.001N^2$ determines the order $O(N^2)$. The best order in data analysis is usually linear $O(N)$. It is hard improve it since usually N entries cannot be worked out by the number of operations less than $O(N)$. The worst order is exponential as $O(\exp N)$. Why they are the best and the worst becomes clear after comparison of N and $\exp N$ for $N = 1000$.

This is one of the possible interpolations. Other interpolations can be constructed by use of the different polynomials by fitting polynomial curves between successive data points. *Mathematica* selects the optimal interpolation in each case. It is possible to calculate the derivatives on the basis of the interpolation. Consider the function $\sin(0.1x)$ at 31 points of the interval $(0, 3)$.

```
In[1]:= T = Table[Sin[i / 10], {i, 0, 30}] // N

Out[1]= {0., 0.0998334, 0.198669, 0.29552, 0.389418, 0.479426, 0.564642,
  0.644218, 0.717356, 0.783327, 0.841471, 0.891207, 0.932039,
  0.963558, 0.98545, 0.997495, 0.999574, 0.991665, 0.973848,
  0.9463, 0.909297, 0.863209, 0.808496, 0.745705, 0.675463,
  0.598472, 0.515501, 0.42738, 0.334988, 0.239249, 0.14112}

In[2]:= f = Interpolation[T]

Out[2]= InterpolatingFunction[{{1., 31.}}, <>]
```

The graph demonstrates typical result of the interpolation. The polynomial interpolation and its derivatives are shown by means of a dashed line, whereas the function $\sin(0.1x)$ by a solid one. One can see that the function and its derivatives are accurately calculated. It is not always that way.

```
In[3]:= Plot[{f[x], Sin[x / 10]}, {x, 1, 30},
  PlotStyle -> {{Thick, Dashed, Black}, {Thick, Black}}]
```

Out[3]=

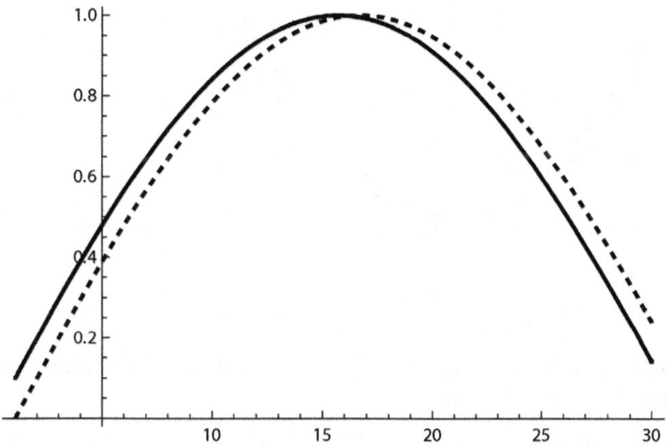

```
In[4]:= Plot[{f'[x], 0.1 Cos[x / 10]}, {x, 1, 30},
    PlotStyle → {{Thick, Dashed, Black}, {Thick, Black}}]
```

Out[4]=

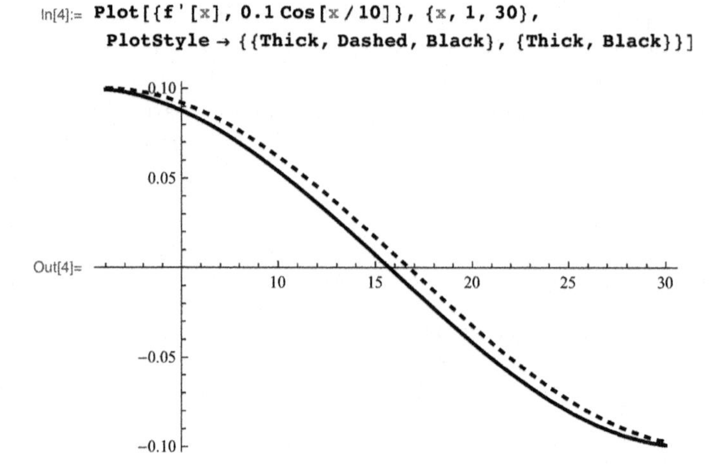


```
In[5]:= Plot[{f''[x], -0.01 Sin[x / 10]}, {x, 1, 30},
    PlotStyle → {{Thick, Dashed, Black}, {Thick, Black}}]
```

Out[5]=

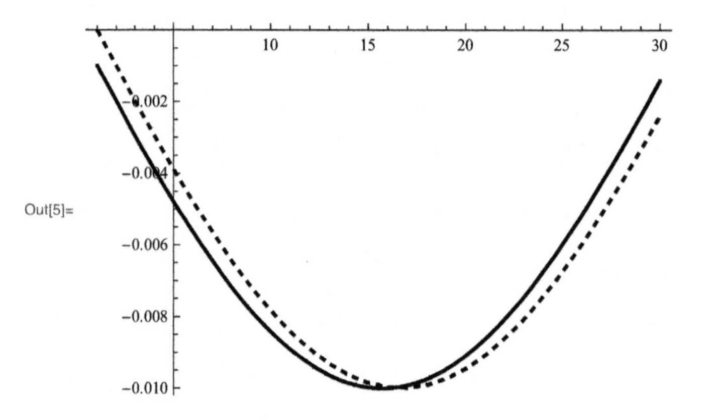

MATLAB Example Box 3.1

The following script uses `interp1` operator using the spline (piecewise polynomial) interpolation.

```
function script08()

    % sample data: 30 evenly spaced points between
    %  0 and 3 and the corresponding values
    x_data   = linspace(0, 3, 30);
    y_data = sin(x_data);

    % query points for interpolated values
    xq = linspace(0, 3, 120);
```

```
% data interpolation
yq = interp1(x_data, y_data, xq,'spline');

%%% PLOTS %%%

% interpolation
subplot(2,2,[1,2]);
plot(xq, yq, 'k--', xq, sin(xq)+0.1, 'k-');
grid on
title('Interpolation');

% 1st derivative based on interpolated points
h = xq(2)-xq(1);    % step size
diff1 = diff(yq)/h;
subplot(2,2,3);
plot(xq(:,1:length(diff1)), diff1, 'k--', xq, ...
    cos(xq)+0.1, 'k')
grid on
title('First derivative')

% 2nd derivative based on interpolated points
diff2 = diff(diff1)/h;
subplot(2,2,4);
plot(xq(:,1:length(diff2)), diff2, 'k--', ...
    xq, -sin(xq)+0.1, 'k')
grid on
title('Second derivative')
```

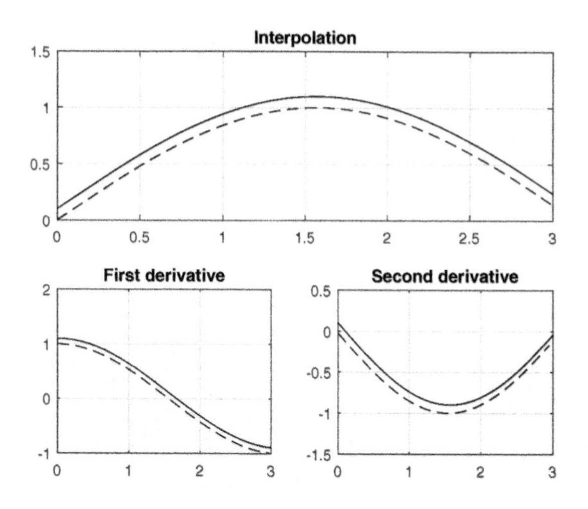

Here, the ellipsis `(...)` continue a long statement to the next line.

If the approximated data (x_i, y_i) $(i = 1, 2, \ldots, n)$ oscillate, it is better to use another approximation instead of interpolation. We proceed with a discussion of an example when interpolation can lead to wrong results. Let data be obtained from not well approximated observations. Let the exact function

$\sin(2\pi x/50)$ be the one with perturbations that are artificially obtained by means of the function $\varepsilon \cos(10x)$ with small $\varepsilon > 0$.

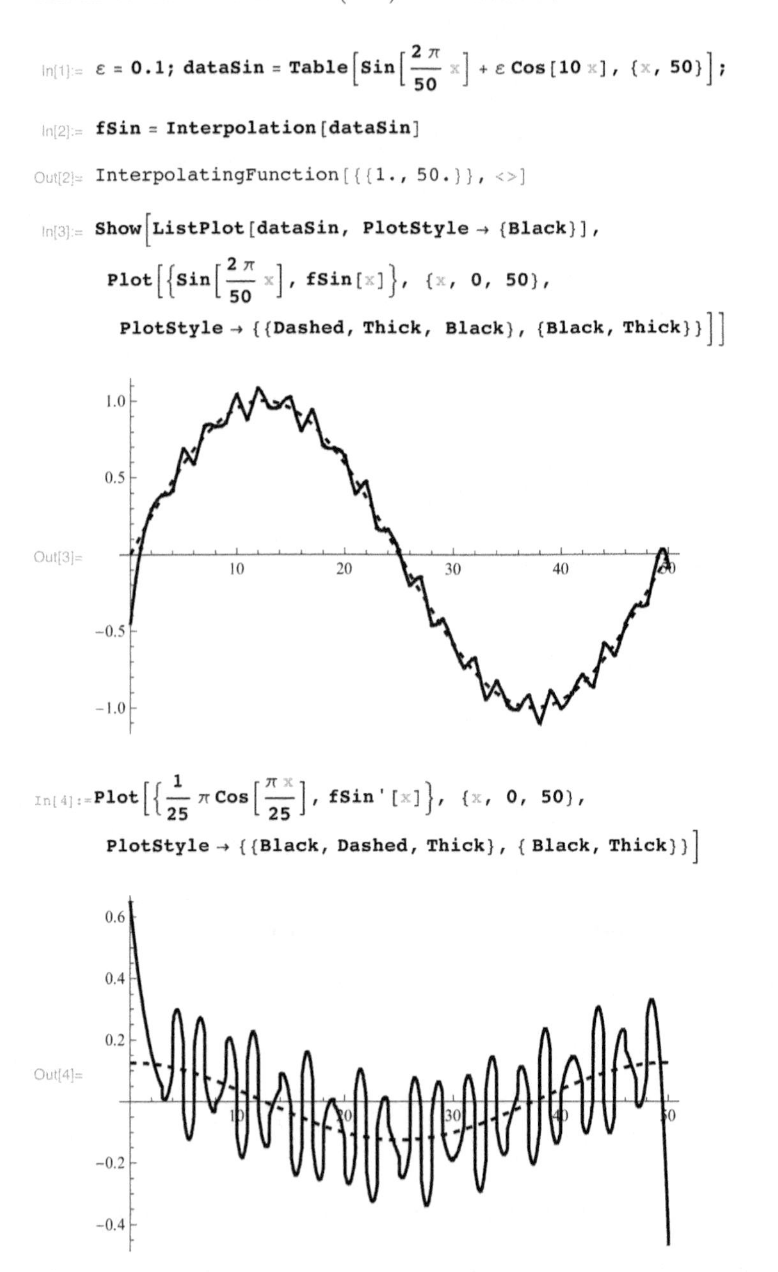

One can see in the first graph that the function is sufficiently well approximated by the interpolation function. But the second graph demonstrates that the derivative of the exact function does not completely coincide with the derivative of the interpolation. It is related to too acute oscillations of the interpolation function what causes significant deviations of the derivative. In such a case it is better to smooth data and afterwards to calculate the derivative (see Sec.5).

MATLAB Example Box 3.2

One can examine the example with perturbed data using the following code.

```
function script09()

    % sample perturbed data
    eps = 0.1;
    x_data = linspace(0, 2*pi, 120);
    y_data = sin(x_data) + eps*cos(100*x_data);

    % query points and interpolation
    xq = linspace(0, 2*pi, 220);
    yq = interp1(x_data, y_data, xq, 'spline');

    % interpolation
    subplot(2,1,1);
    plot(xq, yq, 'k', xq, sin(xq), 'k--');
    grid on
    axis([0 2*pi -1.1 1.1]);
    title('Interpolation');

    % 1st derivative
    subplot(2,1,2);
    h = xq(2)-xq(1);
    diff1 = diff(yq)/h;
    plot(xq(:,1:length(diff1)), diff1, 'k', ...
        xq, cos(xq), 'k--')
    grid on
    axis([0 2*pi -3 3]);
    title('1st derivative');
```

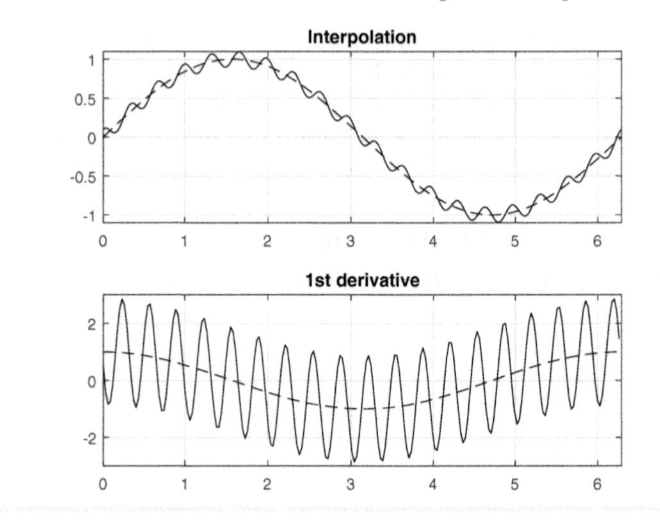

3.3 Integrals

Integrals were discussed in Sec.1.2 and in Sec.2.1. We now proceed with discussion of another important application. It will be shown how the work can be expressed in terms of the integral of forces. Consider a charged particle in the 1D electric field directed along a fixed axis. The motion of the particle can be modeled as a motion of a material point along the axis OX when a force $F(x)$ acting on the point is given at each point x of $[a, b] \subset OX$. Using the principle: *continuous* $\leftrightarrow$ *discrete* we divide the interval $[a, b]$ into small intervals Δx_i $(i = 1, 2, \ldots, n)$. The work done to take the particle on Δx_i is equal to

$$\Delta A_i = F(\xi_i)\Delta x_i, \tag{3.4}$$

where a point ξ_i is arbitrarily chosen on (x_i, x_{i+1}). One can assume that the work ΔA_i is calculated sufficiently well if the interval (x_i, x_{i+1}) is sufficiently short. The work done to bring the particle from a point a to a point b is equal to the sum of works (3.4)

$$A_n = \sum_{i=1}^{n} \Delta A_i = \sum_{i=1}^{n} F(\xi_i)\Delta x_i. \tag{3.5}$$

Let $\delta = \max_i \Delta x_i$ tend to zero. Then, the sum (3.5) tends to a value A, which can be treated as the exact value of the work done to take the particle from a point a to a point b. The limit A of the Riemann sum (3.5) is equal to the definite integral

$$A = \int_a^b F(x)\, \mathrm{d}x. \tag{3.6}$$

The transformation from the simple formula (3.4) into the integral one (3.6) yields the general method to calculate the total value (in this case the work) in the whole interval (a, b). On the other hand the principle *continuous* $\leftrightarrow$ *discrete* can be used in the transformation from the Riemann sum into the discrete rule on the basis of the following

Theorem 3.1 (on average value). *Let a function $F(x)$ be continuous on $[a, b]$. There exists such a point $x_0 \in (a, b)$ that*

$$\frac{1}{b-a}\int_a^b F(x)\ dx = F(x_0). \tag{3.7}$$

This point x_0 changes simultaneously with a and b, i.e. it depends on a and b.

This theorem on average value is also frequently used in continuum mechanics for multiple, surface and other integrals when a, b tends to the fixed $\tilde{x} \in (a, b)$. Then we arrive at the approximate formula (compare to (3.5))

$$\int_a^b F(x)\ \mathrm{d}x \approx (b-a)F(\tilde{x}). \tag{3.8}$$

Let $x = x(t)$ be the equation of the motion of a material point in time t. Then $\mathrm{d}x = \frac{dx}{dt}\mathrm{d}t = v(t)\mathrm{d}t$ where $v(t) = \frac{dx}{dt}$ denotes the velocity of the point. Hence, equation (3.6) can be written in the form

$$A = \int_{t_0}^{t_1} F(x(t))v(t)\ \mathrm{d}t, \tag{3.9}$$

where t_0 is the initial time for which $x(t_0) = a$, t_1 the final time for which $x(t_1) = b$. Multiplication of the force by the velocity yields the capacity $W = Fv$. Therefore, the work from (3.9) can be introduced as the sum of capacities acting during the time interval (t_0, t_1). From the mathematical point of view, equations (3.6) and (3.9) express the substitution rule in the definite integral. Integration affects the spatial variable x in (3.6) and time t in (3.9).

Moment of a particle is expressed by formula $P = mv$, where m denotes the mass of the particle. Then $P = m\frac{dx}{dt}$ and $\frac{dP}{dt} = m\frac{d^2x}{dt^2} = F$ since the force F is the mass (m) multiplied by the acceleration $(a = \frac{d^2x}{dt^2})$. This yields the following formula for the moment change in time (cf. to (3.6) and (3.9))

$$P(t_1) - P(t_0) = \int_{t_0}^{t_1} F\ \mathrm{d}t. \tag{3.10}$$

In order to determine the kinetic energy we consider the force $m\frac{dv}{dt} = F$. Multiply this equation by v:

$$mv\frac{dv}{dt} = Fv \quad \Leftrightarrow \quad \frac{d}{dt}\left(\frac{mv^2}{2}\right) = Fv. \tag{3.11}$$

The kinetic energy K of the moving particle can be introduced by formula

$$K = \frac{mv^2}{2}. \tag{3.12}$$

Hence, $\frac{dK}{dt} = Fv$. It follows from (3.9) that

$$A = \int_{t_0}^{t_1} Fv \, dt = \int_{t_0}^{t_1} \frac{dK}{dt} \, dt = K(t_1) - K(t_0). \tag{3.13}$$

Equation (3.13) expresses *Law of Conservation of Energy: the work done to bring the particle is equal to the change of the kinetic energy.*

3.4 Potential energy

Potential energy is defined up to an arbitrary additive constant u_0 as the integral of force:

$$u(x) = -\int_a^x F(x) \, dx + u_0. \tag{3.14}$$

It follows from (3.14) that $u(a) = u_0$. Differentiation of (3.14) in x yields

$$F = -\frac{du}{dx} \quad \Leftrightarrow \quad m\frac{dv}{dt} = -\frac{du}{dx}. \tag{3.15}$$

In the latter equation, the derivative in t is taken in the left-hand part and the derivative in x in the right-hand part. We try to transform $\frac{dv}{dt}$ in such a way that the derivative in t will disappear. It follows from the chain rule formula that

$$\frac{dv}{dt} = \frac{dv}{dx}\frac{dx}{dt} = v\frac{dv}{dx}.$$

This yields

$$m\frac{dv}{dt} = mv\frac{dv}{dx} = \frac{d}{dx}\left(\frac{mv^2}{2}\right).$$

Use equation (3.15)

$$\frac{d}{dx}\left(\frac{mv^2}{2}\right) = -\frac{du}{dx}.$$

It can be written as the energy conservation law

$$\frac{mv^2(x)}{2} + u(x) = E, \tag{3.16}$$

where the constant E is *the total energy* of a particle. This constant E can be determined through the initial condition

$$E = \frac{mv_0^2}{2} + u_0,$$

where v_0 stands for the velocity of particle at the initial time t_0.

Sometimes it is convenient to consider the variables x and v as independent variables without using time. These two variables form *phase space*. Then, equation (3.16) becomes

$$\frac{mv^2}{2} + u(x) = E. \tag{3.17}$$

It can be considered as an equation of the curve in the phase space of the variables x and v. Different E yields different curves. Each curve corresponds to a motion of a particle with the fixed energy E. Advantages of this description will be demonstrated by means of examples in Secs 4.6-4.7.

The energy conservation law (3.16) can be used to describe the standard motion of a particle in time. Equation (3.16) can be considered as an ordinary differential equation on the trajectory $x(t)$

$$\frac{m}{2}\left(\frac{dx}{dt}\right)^2 + u(x) = E. \tag{3.18}$$

The initial condition for equation (3.18) has the form

$$x(t_0) = a. \tag{3.19}$$

An ordinary differential equation with energy is frequently simplified by differentiation in the spatial variable x

$$m\frac{d^2x}{dt^2} + u'(x) = 0. \tag{3.20}$$

This equation can be also deduced from the equilibrium of forces acting on the material point.

Concluding remarks and further reading. We recommend the books [38, 39, 40, 49] devoted to the classical mechanics.

Exercises

1. Let a particle move in the space by the trajectory $\mathbf{S}(t) = (4\sin t, 3\cos 2t, \frac{1-t^2}{10})$. Calculate its velocity and acceleration. Display the trajectory online.

   Use the *Mathematica* operators **`Derivative`**, **`ParametricPlot3D`**, **`Animate`**, **`Manipulate`**.

   Display the trajectory in MATLAB using `fplot3` operator. For creating animation refer to Example Box 8.1 on page 175.

2. Interpolate the data $\{\{0, 0.22\}, \{0.5, -0.99\}, \{1, -0.0088\}, \{1.5, 0.99\}, \{2, 0.017\}, \{2.5, -0.99\}, \{3, -0.026\}\}$. Show the result on graph.

3. Calculate the integral

$$a) \int_0^{\pi} \sin nx \arctan\left(\frac{\tan\frac{x}{2}}{\tan\frac{a}{2}}\right) dx$$

for various natural n. Take, for instance, $n = 6, 8, 9$.

Hint: Use the *Mathematica* operators **Simplify** and **Assumptions** with the declarations that n belongs to integers and $0 < a < 0.1$. Try the declaration $a > 0$.

Calculate the integrals

$$b) \int_{-\infty}^{+\infty} \exp(-x^2) dx;$$

$$c) \int_0^{\infty} x^n \exp(-ax^2)\, dx \quad \text{for even and odd } n, \text{ and } a > 0;$$

$$d) \int_{-\infty}^{\infty} \frac{\sin(\alpha x)}{x}\, dx \quad \text{for } \alpha > 0;$$

$$e) \int_0^{a} \exp(-x^2)\, dx.$$

4. Let a particle move along the x-axis in the force field $F(x) = -x^3$. Display the phase space of the motion by means of equation (3.17).

Chapter 4

Ordinary differential equations and their applications

4.1 Principle of transition for ODE 58
4.2 Radioactive decay 59
4.3 Logistic differential equation and its modifications 61
4.3.1 Logistic differential equation 61
4.3.2 Modified logistic equation 61
4.3.3 Stability analysis 64
4.3.4 Bifurcation 66
4.4 Time delay 68
4.5 Approximate solution to differential equations 68
4.5.1 Taylor approximations 69
4.5.2 Padé approximations 71
4.6 Harmonic oscillation 73
4.6.1 Simple harmonic motion 73
4.6.2 Harmonic oscillator with friction and exterior forces ... 76
4.6.3 Resonance 80
4.7 Lotka-Volterra model 83
4.8 Linearization 86
Exercises 88

An ordinary differential equation (ODE) is an equation containing an unknown differentiable function $y(t)$ and its derivatives. For instance, equation (1.1) is a differential equation of second order, since the second derivative is the maximal order derivative contained in (1.1). *The general solution* (1.2) of equation (1.1) contains two arbitrary constants C_1 and C_2. The corresponding Cauchy's problem consists of the differential equation (1.1) and two initial conditions (1.3)-(1.4). Use of these conditions determines the constants C_1, C_2 and yields *the particular solution* (1.4) of the differential equation (1.5).

A course of ordinary differential equations can be conditionally divided into the following three branches. The first one deals with finding solution of an ODE or a corresponding Cauchy's problem. There are various methods to solve ODE. The main task is to determine which ODE class a given equation belongs to. When the class is identified, the corresponding set of manipulations is applied to find a solution. Not each equation belongs to such a class with the prescribed rules to solve it. Then, one can try to solve an ODE inviting a new method, i.e., to introduce a new class of equations. However, not each equation can be solved, i.e., its solution cannot be written in terms of elementary (well-known) functions. Such a situation took place at the beginning of ODE (Gottfried Wilhelm von Leibniz, 1675) when a researcher came across an unsolvable equation and solved it by a "linguistic method", when an unknown function was defined as a solution of the given ODE. Thus, Friedrich Wilhelm Bessel (lived in 1784-1846) studied the following equation

$$t^2y''(t) + ty'(t)(t^2 - n^2)y(t) = 0, \tag{4.1}$$

where n is a natural number. He noted that it is impossible to express solutions of this equation by elementary functions. Later this equation (4.1) was called by his name and its solutions by Bessel's functions. So, if Smith cannot solve an equation, he can introduce new Smith's functions as functions satisfying this equation.

Frequently, in Mathematical Modeling we identify a differential equation with a mathematical model. For instance, we say that equation (1.1) models a projectile or equation (1.1) is a model of the projectile. This stresses the exceptional role of differential equations in Mathematical Modeling.

The best way to solve an equation is provided by *Mathematica* and by MATLAB. The operator **DSolve** solves an ordinary differential equation including eventual symbolic form solution. The operator **NDSolve** finds a numerical solution.

The second branch of the ODE theory is devoted to qualitative investigations (existence and uniqueness of solutions, stability, relations to algebra and geometry).

The third branch is related to applications to mechanics and other sciences (biology, economy, chemistry etc). It also includes numerical methods for ODE which cannot be solved in terms of elementary and special functions. In this book, we pay attention to applications.

4.1 Principle of transition for ODE

Principle of transition: *continuous* $\leftrightarrow$ *discrete* (formulated on page 19) helps to develop a model of processes in time by the following lines. Let an unknown value be presented as a function $y(t)$ depending on time t. Let Δt denote a sufficiently small increment of time. The increment of $y(t)$ is calculated by formula

$$\Delta y = y(t + \Delta t) - y(t).$$

Usually, Δy expresses a balance of substance which is proportional to Δt. Let it have the form

$$\Delta y = f(t, y)\Delta t, \tag{4.2}$$

where $f(y,t)$ is a given function. Dividing (4.2) by Δt and calculating the limit $\Delta t \to 0$ we arrive at the ODE

$$y'(t) = f(t, y(t)). \tag{4.3}$$

Example 4.1. Let a reservoir with two holes contain 10 litre (l) of pure water at the beginning. Let 1.5 l per minute of 1% salt water enter into the reservoir through the first hole, mix within it and let 1.5 l per minute of mixture go out through the second hole. Determine the salt concentration in the reservoir after 10 min.

Let $y(t)$ denote the volume of salt at time t. It is a dimensional value measured in l. Then, the salt concentration at time t is equal to $0.1y(t)$. It is a dimensionless value as ratio of $y(t)$ l to 10 l. We have

$$y(0) = 0. \tag{4.4}$$

Calculate the balance of salt from time t to $t+\Delta t$. During the time interval Δt the volume 1.5 Δt of solution enters into the reservoir. This volume contains $0.01 \cdot 1.5\ \Delta t = 0.015\ \Delta t$ of salt. The volume 1.5 Δt of mixture goes out from the reservoir. Hence, this mixture contains $0.1y(t) \cdot 1.5\Delta t = 0.15y(t)\Delta t$ of salt. Therefore, 0.015 Δt of salt enters and $0.15y(t)\ \Delta t$ goes out during the time Δt. This yields the balance of the salt

$$\Delta y = 0.015\ \ \Delta t - 0.15y(t)\ \Delta t. \tag{4.5}$$

Divide (4.5) by Δt and pass to the limit $\Delta t \to 0$. We get the differential equation

$$y'(t) = 0.015 - 0.15y(t). \tag{4.6}$$

Cauchy's problem (4.6), (4.4) has the unique solution which can be found by the operator **`DSolve`**

$$y(t) = 0.1 - 0.1\exp(-0.15t). \tag{4.7}$$

The volume of salt at time $t = 10$ is equal to $y(10) = 0.0777$, hence the salt concentration is equal to 0.00777.

4.2 Radioactive decay

The law of the radioactive decay maintains that the mass decay Δm during the time Δt is proportional to the total mass of the radioactive sample m. Let the coefficient of the proportionality be equal to $-k$ where k is a positive constant. Then, the law can be written in the form

$$\Delta m = -km\Delta t. \tag{4.8}$$

The minus sign shows that the mass decreases. Divide (4.8) by Δt and take the limit as $\Delta t \to 0$. According to the definition of derivatives we obtain the following ODE

$$m'(t) = -km(t) \iff \frac{dm}{dt} = -km, \tag{4.9}$$

where the unknown function is $m(t)$. Let the initial condition be given

$$m(0) = m_0. \tag{4.10}$$

We arrive at the Cauchy problem (4.9)-(4.10). Separation of variables in the ODE[1] (4.9) yields

$$\frac{dm}{m} = -kdt \iff \int \frac{dm}{m} = -k \int dt \iff \ln m = -kt + C, \tag{4.11}$$

where C is an arbitrary integration constant. We have

$$m(t) = ce^{-kt}, \tag{4.12}$$

where the constant $c = e^C$ is introduced. Using the initial condition (4.10) we get

$$m(t) = m_0 e^{-kt}. \tag{4.13}$$

The function (4.13) expresses the exponential law of radioactive decay.

The law (4.8) can be referred to by another process. For instance, let us locate a sum of money in a bank. Let it be the initial sum m_0. The bank pays an interest rate k which is usually proportional to the amount $m(t)$ in time t and to time Δt. Then,

$$\Delta m = km\Delta t. \tag{4.14}$$

Continuous approximation of (4.14) yields the differential equation

$$\frac{dm}{dt} = km. \tag{4.15}$$

Its solution yields the exponential law similar to (4.13)

$$m(t) = m_0 e^{kt}. \tag{4.16}$$

Of course, the continuous formula can be applied for a long time.

The law (4.14) can be treated as the Malthus (1798) population law when the increment of population Δm during time Δt is proportional to the population size $m(t)$.

---

[1] see also Sec.8.5

4.3 Logistic differential equation and its modifications

4.3.1 Logistic differential equation

Equations (4.9) and (4.16) can describe a biologic population. These equations do not take into account that resources (food) necessary for its growth can be restricted. The coefficient k in such a model can be reduced with the growth of the population denoted here as $x(t)$. In particular, it can be introduced as $k[1 - ax(t)]$. Then, instead of (4.9) we arrive at the ODE

$$x'(t) = k\, x(t)[1 - ax(t)]. \tag{4.17}$$

Its solution can be obtained by means of the operator **DSolve** and it has the form

$$x(t) = \frac{e^{kt}}{ae^{kt} - C}, \tag{4.18}$$

where C is an arbitrary constant. The initial condition

$$x(0) = x_0. \tag{4.19}$$

yields the logistic curve determined by the function

$$x(t) = \frac{x_0 e^{kt}}{ax_0\left(e^{kt} - 1\right) + 1}. \tag{4.20}$$

4.3.2 Modified logistic equation

In the previous subsection, we used the standard procedure to modify the model (4.15) and to develop the model (4.17). It is the procedure of steps (see page 12) taken to create a mathematical model. We proceed with this line and explain development of the insect outbreak model due to Ludwig et al. [48, v. 1, page 7]. First, try to study the budworm population dynamics by equation (4.17) where $x(t)$ denotes the population density of the spruce budworms, k the linear birth rate and $b = a^{-1}$ the carrying capacity which is related to the density of food. Experimental observations demonstrate that equation (4.17) has to be modified by addition of the external predation term $-p[x(t)]$ which takes into account birds eating insects

$$x'(t) = k\, x(t)\left[1 - \frac{x(t)}{b}\right] - p[x(t)]. \tag{4.21}$$

The function $p = p(x)$ is worthy of notice. Its introduction is a separate task of mathematical modeling. The dimensional analysis (see Sec.1.4, page 22) shows that p means the number of birds per area×time, for instance per $km^2 \times day$.

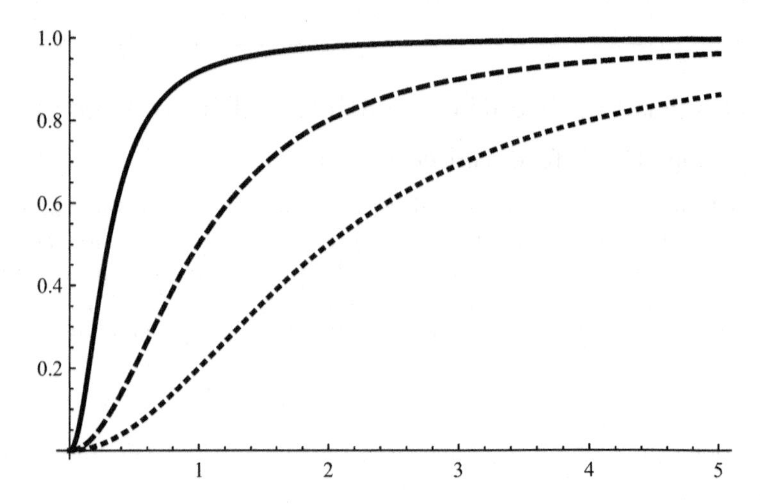

FIGURE 4.1: Graph of (4.22) for $\alpha = 1$ and $\beta = 0.3$ (solid line), $\beta = 1$ (dashed line) and $\beta = 2$ (dotted line).

Observations and logical reasoning suggest that $p(x)$ has different regimes on three different segments. Namely, $p(x)$ should be small on $0 \leq x < x_1$ ($p(0) = 0$ yields an unstable regime of fatal meetings), approximately linear on $x_1 \leq x < x_2$ (a stable regime when predators (birds) chase insects and the number of their meetings is proportional to x) and constant on $x \geq x_2$ (too many birds leading to a saturation regime). Following the principle of the simplest model stated on page 12, we try to find such a simple function and suggest that it is

$$p(x) = \frac{\alpha x^2}{\beta^2 + x^2}. \tag{4.22}$$

$p(x)$ is displayed in Fig.4.1. Numerical solution of (4.21) can be found by the following computations

```
In[1]:= p[x_, α_, β_] = (α x^2)/(β^2 + x^2);

In[2]:= s = NDSolve[{x'[t] == x[t] (1 - x[t]) + p[x[t], 1, 1], x[0] == 1},
          x, {t, 0, 4}]

Out[2]= {{x → InterpolatingFunction[{{0., 4.}}, <>]}}
```

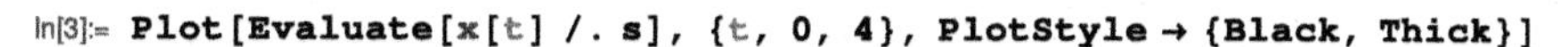

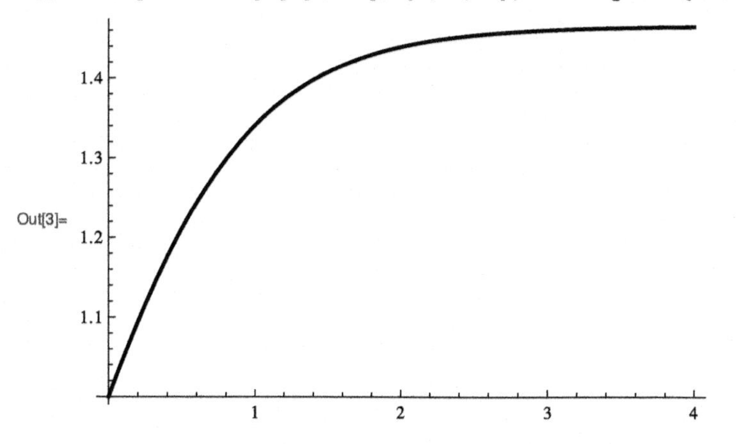

In order to reduce the number of parameters we introduce the new unknown $u = \frac{x}{\beta}$ and the parameters $r = \frac{k}{\alpha\beta}$, $q = \frac{b}{\beta}$. Then, (4.21) becomes

$$u' = f(u, r, q), \tag{4.23}$$

where

$$f(u, r, q) = u\left[r\left(1 - \frac{u}{q}\right) - \frac{u}{1 + u^2}\right] \tag{4.24}$$

MATLAB Example Box 4.1

One can easily visualize the solution of the logistic equation using `ode45` operator introduced in Example Box 1.1 on page 7.

```
function script10()

    t_interval = [0 4]; % time interval
    y0 = 1;             % initial condition

    % function definitions
    p = @(x, a, b) (a*x^2) / (b^2 + x^2);
    f = @(t, x) x * (1-x) + p(x, 1, 1);

    % solution of the ODE
    [T, X] = ode45(f, t_interval, y0);

    % plot
    plot(T, X)
    grid on
```

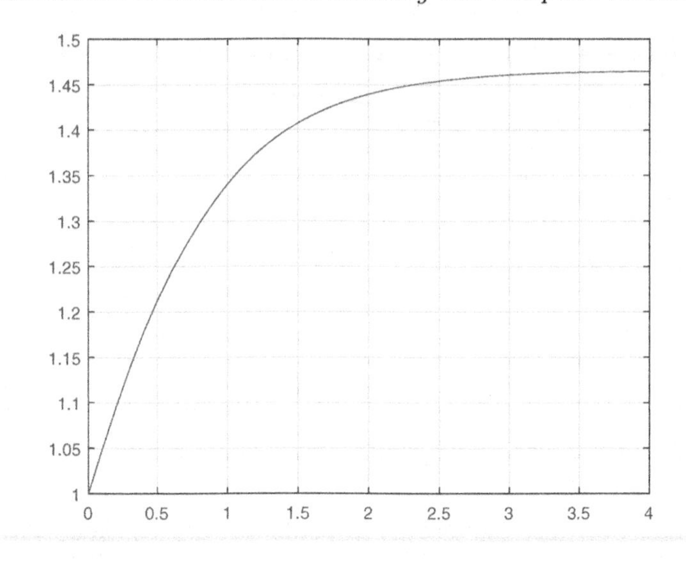

4.3.3 Stability analysis

The last graph from the previous subsection, in particular, demonstrates that the population x, hence the value u, tends to a constant value with time. The steady state $x(t) = constant$ for equation (4.21) takes place when $x'(t) \equiv 0$ that is equivalent to $u'(t) \equiv 0$. Excluding the trivial case $u = 0$ we arrive at the algebraic equation

$$r\left(1 - \frac{u}{q}\right) - \frac{u}{1+u^2} = 0. \tag{4.25}$$

The latter equation is equivalent to a cubic equation. Three roots of (4.25) $u_n(r,q)$ $(n = 1, 2, 3)$ can be explicitly written by use of the operator **Solve** or approximately by the operator **FindRoot**. Examples when equation (4.25) has one or three positive roots for different parameters r and q are shown in Fig.4.2. The case of two roots will be discussed in the next section. It is worth noting that a cubic equation with real coefficients always has at least one real root (explain why?).

The behavior of the function (4.24) determines the *equilibrium states of the models* (4.21) and (4.23). The general scheme is based on the Taylor approximation

$$f(u_1 + v, r, q) = f(u_1, r, q) + f'(u_1, r, q)v + O(v^2), \tag{4.26}$$

where f' stands for the derivative in the first variable u. Let u_1 be an equilibrium state, i.e., the constant u_1 is a root of equation $f(u_1, r, q) = 0$. Let $v = v(t)$ denote a small perturbation of the constant solution $u(t) = u_1$. Then, (4.23) and (4.26) yield the approximate simple ODE

$$v' \approx \lambda v, \tag{4.27}$$

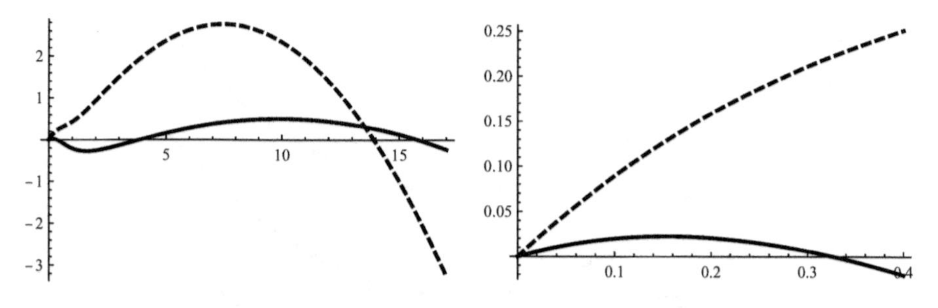

FIGURE 4.2: Graphs of the functions $f(u, 0.3, 20)$ (solid line) and $f(u, 1, 15)$ (dashed line). One can analyze the roots of the functions for small u on the right figure. Application of **DSolve** gives three positive roots $u_1 = 0.326574$, $u_2 = 3.87692$, $u_3 = 15.7965$ for equation $f(u, 0.3, 20) = 0$ and one root $u_1 = 13.9286$ for equation $f(u, 1, 15) = 0$.

where the constant $\lambda = f'(u_1, r, q)$. The solution of (4.27) has the form

$$v(t) = v(0)\exp(\lambda t). \tag{4.28}$$

One can see that the disturbance $v(t)$ is unbounded for all $t > 0$ if $\lambda > 0$ and bounded in the opposite case. This implies that the equilibrium state $u = u_1$ is unstable if $\lambda > 0$ and stable if $\lambda < 0$. Usually, the case $\lambda = 0$ requires a separate investigation.

Consider numerical examples. The function $f(u, 0.3, 20)$ has the root $u_1 = 0.326574$. Using the operator **Series** we find that $f(v, 0.3, 20) \approx -0.243121v$ in the vicinity of the point $v = u - u_1$, hence $v' \approx -0.243121v$. Instead of **Series** one can calculate $f'(u_1, 0.3, 20) = -0.243121$. Solution to the latter ODE has the form

$$v(t) \approx v(0)e^{-0.243121t}.$$

This implies that $v(t)$ tends to the equilibrium state for $t \to +\infty$. Therefore,

$$u(t) \approx u_1 + [u(0) + u_1]e^{-0.243121t}$$

also tends to its equilibrium state u_1. In this case, the equilibrium state $u = u_1$ is stable in time and $u(t)$ approaches the constant value u_1 if the initial value $u(0)$ is close to u_1 (u_1 is an attractive point).

Similar investigation near the point $u_2 = 3.87692$ leads to the result

$$u(t) \approx u_2 + [u(0) + u_2]e^{0.153519t}.$$

One can see that the latter function is unbounded as $t \to +\infty$. This means that the equilibrium state $u = u_2$ is not stable and u_2 is a repulsive point.

Actually, we have performed the linearization and stability analysis of the model (4.21) near the points u_1 and u_2 presented from the general point of view in Sec.4.8. We have shown that a graph like the one on page 63 in general can be unstable, e.g., it can be transformed to another form after small perturbations similar to the ball on the hill from Fig.1.6.

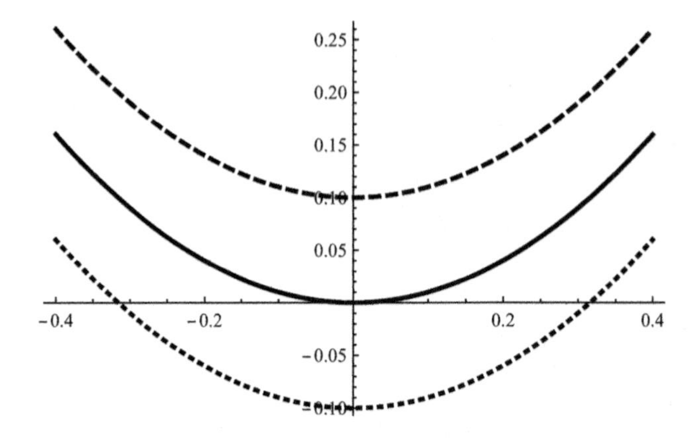

FIGURE 4.3: Graphs of $x(t) = t^2 - \varepsilon$ for $\varepsilon = 0$ (solid line), for $\varepsilon > 0$ (dashed line) and for $\varepsilon < 0$ (dotted line).

4.3.4 Bifurcation

The above investigation can be developed to the bifurcation analysis shortly presented below. First, we discuss a simple model, and then come back to our equation (4.25). Consider the following equation

$$t^2 - \varepsilon = 0 \tag{4.29}$$

with a small parameter $|\varepsilon|$. Fig.4.3 demonstrates that small perturbations of ε near zero change the number of roots of equation (4.29). The point $\varepsilon = 0$ is called the *bifurcation point* for equation (4.29).

The cubic equation

$$r(1+u^2)\left(1-\frac{u}{q}\right) - u = 0 \tag{4.30}$$

is equivalent to (4.25). A polynomial equation $P(u)$ has at least a double root at $u = u_0$ if $P(u_0) = 0$ and $P'(u_0) = 0$. Application of this double condition root to (4.30) yields the system of algebraic equations

$$\begin{cases} r(1+u^2)\left(1-\frac{u}{q}\right) - u = 0 \\ 3ru^2 - 2rqu + r + q = 0 \end{cases} \tag{4.31}$$

Application of the operator **Reduce** yields

```
In[1]:= Reduce[{r (1 - u/q) (1 + u²) - u == 0, q + r - 2 q r u + 3 r u² == 0},
  {u, r, q}]

Out[1]= 1 + u² ≠ 0 && r == 2 u³/(1 + u²)² && -1 + u² ≠ 0 && q == 2 u³/(-1 + u²) && u ≠ 0
```

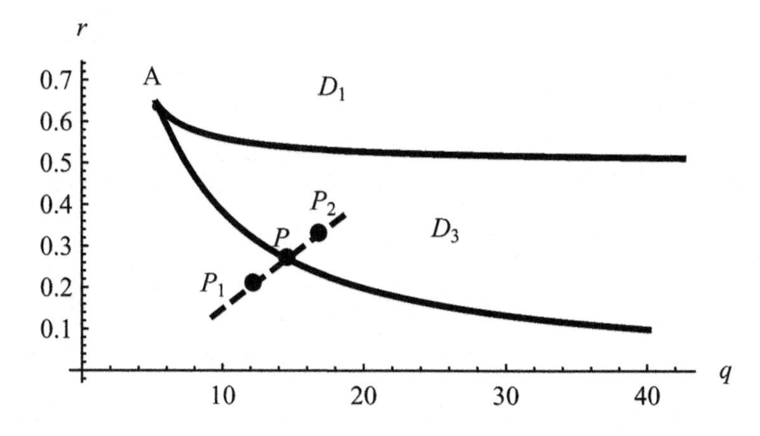

FIGURE 4.4: Double root curve (4.32). Equation (4.30) has three positive roots for q and r in the domain D_3 and one root in the domain D_1.

The obtained result can be treated in the following way. Equations

$$\begin{cases} r(u) = \frac{2u^3}{(u^2+1)^2} \\ q(u) = \frac{2u^3}{u^2-1} \end{cases} \tag{4.32}$$

are parametric equations of the double root curve shown in Fig.4.4. Let a point P be fixed on the curve (4.32). Consider small perturbations of P, the point P_1 in the one root domain and P_2 in the three roots domain that corresponds to small perturbations of the parameters p and q. Such a perturbation can drastically change the behavior of the model when the point (q, r) moves from P_1 to P_2 through P. Because u passes through the branch point P and takes one of three possibilities in the domain D_3, i.e., the model jumps to one of the three possible regimes. This point P is the bifurcation point of the model and the double root curve in Fig4.4 is the bifurcation curve.

The boundary of D_3 contains the wonderful cusp point $A(3\sqrt{3}, \frac{3}{8}\sqrt{3})$ where the derivatives of functions (4.32) vanish at $u = \sqrt{3}$. This surface $u = u(p, q)$ defined by (4.25) near A models the cusp catastrophe [7] (see the source code in `http://demonstrations.wolfram.com/DegenerateCriticalPointsAndCatastrophesFoldCatastrophe/`).

Remark 4.1. Beginning from Sec.4.2 we permanently improve and simultaneously complicate the mathematical model by taking into account new features of the processes discussed. We come to equation (4.21) from the simple equation (4.9) by addition the further parameters. It can be seen that it is possible to improve the model infinitely by adding new parameters and perhaps to develop the perfect model. However, such a romantic passage has to be ended by the quote due to John von Neumann: "With four parameters I can fit an elephant, and with five I can make him wiggle his trunk".

4.4 Time delay

A time delay is a natural parameter for population dynamics to take account of the time to reach maturity. Consider the modified equation

$$x'(t) = -kx(t - t_0), \tag{4.33}$$

where t_0 is a given time delay. We are looking for solutions of (4.33) in the form

$$x(t) = ce^{\lambda t}, \tag{4.34}$$

where λ is an unknown complex number. The real constant c is determined by the initial condition $x(0) = c$. Substitution of (4.34) into (4.33) yields the transcendental equation on λ

$$\lambda = -ke^{-\lambda t_0}. \tag{4.35}$$

Using **FindRoot** one can check that equation (4.35) does not always have real roots λ. It can be established graphically by analysis of the joint points of the plots $y(\lambda) = \lambda$ and $y(\lambda) = -ke^{-\lambda t_0}$. The complex roots $\lambda = \mu + i\omega$ of (4.35) yield the solutions

$$x(t) = ce^{\mu t}\cos\omega t, \quad x(t) = ce^{\mu t}\sin\omega t. \tag{4.36}$$

4.5 Approximate solution to differential equations

Consider Cauchy's problem

$$x'(t) = x^m(t) + 1, \quad x(0) = 0. \tag{4.37}$$

Application of the operator **DSolve** gives $x(t) = \tan x$ for $m = 2$. Attempts to solve (4.37) for $m = 3$ fail. Application of **NDSolve** gives the function $x(t)$ in a numeric form. Such a form is not always acceptable. We shall try to solve the problem (4.37) for $m = 3$ by means of polynomial and rational approximations that will give an approximate solution in an analytical form. It follows from the theory of ordinary differential equations [61] that the problem (4.37) for $m = 3$ has a unique solution for sufficiently small $t > 0$. Equation (4.37) implies that the continuous function $x(t)$ can be differentiated infinitely many times (infinitely differentiable function). It follows from equation

$$x'(t) = x^3(t) + 1 \tag{4.38}$$

that the derivative $x'(t)$ exists and is a continuous function, since the right-hand side is a continuous function. Differentiation of (4.38) implies that $x''(t)$ is a continuous function. This scheme holds for any derivative $x^{(n)}(t)$. As we will see later that difficulties arise when we will try to continue a solution to large t.

4.5.1 Taylor approximations

In the present section, the power series method is applied to ordinary differential equations. We are looking for an approximate expression for the function $x(t)$ in the polynomial form

$$x(t) \approx \sum_{n=0}^{N} c_n t^n. \tag{4.39}$$

In order to determine the coefficients c_n we use the Taylor formula

$$c_n = \frac{x^{(n)}(0)}{n!}. \tag{4.40}$$

The derivative of an unknown function at zero is needed to compute c_n. Differentiation of (4.38) and substitution of zero into the result gives the coefficients c_n. Do it with a computer. Introduce the derivative of order n as **Der[n]** and **c[n]** as its value at zero. We have

```
In[1]:= Der[1] = x[t]^3 + 1
        c[0] = x[0] = 0
        c[1] = x'[0] = Der[1] /. t → 0

Out[1]= 1 + x[t]^3

Out[2]= 0

Out[3]= 1
```

Let us look at the computation of the high order derivatives:

```
In[4]:= x''[t]
        % // FullForm

Out[4]= x''[t]

Out[5]//FullForm=
        Derivative[2][x][t]
```

Differentiate (4.38) and substitute zero into the result

```
In[6]:= Der[2] = D[Der[1], t]
        c[2] = Derivative[2][x][0] = % /. t → 0

Out[6]= 3 x[t]^2 x'[t]

Out[7]= 0
```

Analogously, introduce

```
In[8]:= Der[3] = D[Der[2], t]
        c[3] = Derivative[3][x][0] = % /. t → 0

Out[8]= 6 x[t] x'[t]^2 + 3 x[t]^2 x''[t]

Out[9]= 0
```

Now, we can compute derivatives of any order, for instance up to the 20th order (for the sake of briefness the whole result Out is not completely written)

```
In[10]:= N = 20; Do[Print[
           "x"^n, "(t):= ",
           Der[n] = D[Der[n - 1], t], " ↦ ",
           c[n] = Derivative[n][x][0] = Der[n] /. t → 0
          ], {n, 4, N}]

x^4 (t):= 6 x'[t]^3 + 18 x[t] x'[t] x''[t] + 3 x[t]^2 x^(3)[t]  ↦ 6
```

The Taylor series is computed by formula

$$\text{In[11]:= Taylor[t\_]} = \sum_{n=0}^{N} \frac{c[n]}{n!} t^n$$

$$\text{Out[11]=}\; t + \frac{t^4}{4} + \frac{3\,t^7}{28} + \frac{57\,t^{10}}{1120} + \frac{737\,t^{13}}{29\,120} + \frac{42\,153\,t^{16}}{3\,261\,440} + \frac{416\,181\,t^{19}}{61\,967\,360}$$

Construct a graph of the obtained polynomial **Taylor[t]**.

```
In[12]:= Plot[Taylor[t ], {t, 0, 5} , PlotStyle → {Thick, Black, Dashed},
          PlotRange → {-1, 5}]
```

Out[12]=

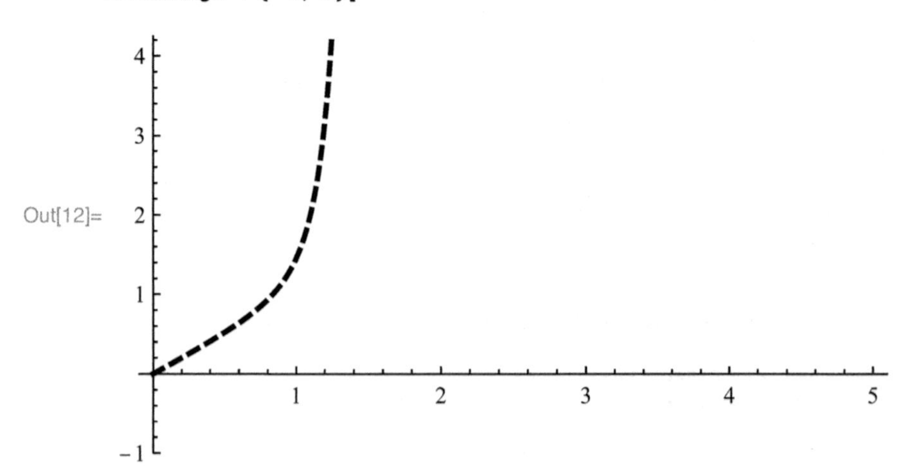

4.5.2 Padé approximations

One can notice that the Taylor polynomials, as any polynomial, do not have singular points, i.e., they do not possess vertical asymptotes. This implies that a polynomial cannot approximate a function with singularities sufficiently well. Let us try to find a rational approximation for such a function. *Padé approximations* are the best rational approximations in some sense. Padé approximations are constructed by means of a given Taylor polynomial (4.39) by the following formula

$$x(t) \approx c_0 + c_1 x + \cdots + c_N x^N = \frac{a_0 + a_1 x + \cdots + a_P x^P}{1 + b_1 x + \cdots + b_Q x^Q} + O(x^{N+1}), \quad (4.41)$$

where $P + Q = N$. The coefficients a_n and b_n are uniquely determined in terms of c_n by equating coefficients of x^n $(n = 0, 1, 2, \ldots, N)$ in the formula following from (4.41)

$$(c_0+c_1x+\cdots+c_Nx^N)(1+b_1x+\cdots+b_Qx^Q) = a_0+a_1x+\cdots+a_Px^P+O(x^{N+Q+1}) \quad (4.42)$$

The rational function from the right-hand side of (4.41) is called $[P/Q]$ *Padé approximation.* Padé approximation is determined by the operator

```
In[13]:= Pade[t_] = PadeApproximant[Taylor[t], {t, 0, {10, 10}}]
```

$$\text{Out[13]=} \quad \frac{t - \frac{48150853\, t^4}{65850884} + \frac{494401599\, t^7}{3687649504} - \frac{378204391\, t^{10}}{87162624640}}{1 - \frac{32306787\, t^3}{32925442} + \frac{143412333\, t^6}{526807072} - \frac{870559471\, t^9}{47939443552}}$$

Advantages of the Padé approximation are demonstrated by the following graph

```
In[14]:= Plot[{Pade[t], Taylor[t]}, {t, 0, 10}, PlotRange → {-5, 5},
  PlotStyle →
   {{AbsoluteThickness[1.5], AbsoluteDashing[{15, 8}], Black},
    {AbsoluteThickness[4], AbsoluteDashing[{10, 9}], Black}}]
```

Out[14]=

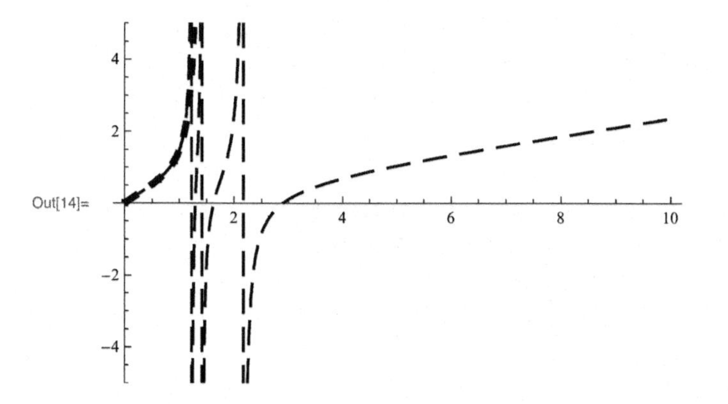

One can see that the Taylor polynomial does not capture singularities. However, one can mistrust this figure, since we do not know the true function satisfying (4.38). In order to verify the results one can use a numerical approximation obtained by the operator **NDSolve** at least for small $t > 0$.

Effectiveness of the Padé approximation can be confirmed by the problem (4.37) with $m = 2$. Its solution is $x(t) = \tan t$. Now, we formally repeat manipulation leading to Padé approximation of the simple problem under consideration

```
In[1]:= Der[1] = x[t]^2 + 1
        c[0] = x[0] = 0
        c[1] = x'[0] = Der[1] /. t → 0

Out[1]= 1 + x[t]^2

Out[2]= 0

Out[3]= 1

In[4]:= Der[2] = D[Der[1], t]
        c[2] = Derivative[2][x][0] = % /. t → 0

Out[4]= 2 x[t] x'[t]

Out[5]= 0

In[6]:= Der[3] = D[Der[2], t]
        c[3] = Derivative[3][x][0] = % /. t → 0

Out[6]= 2 x'[t]^2 + 2 x[t] x''[t]

Out[7]= 2

In[8]:= N = 20; Do[Print[
          "x"^n, "(t):= ",
          Der[n] = D[Der[n - 1], t], " ↦ ",
          c[n] = Derivative[n][x][0] = Der[n] /. t → 0
         ], {n, 4, N}]

        x^4 (t):= 6 x'[t] x''[t] + 2 x[t] x^(3)[t] ↦ 0
```

In[9]:= **Taylor[t_] =** $\sum_{n=0}^{N} \frac{c[n]}{n!} t^n$

Out[9]= $t + \frac{t^3}{3} + \frac{2 t^5}{15} + \frac{17 t^7}{315} + \frac{62 t^9}{2835} + \frac{1382 t^{11}}{155\,925} + \frac{21\,844 t^{13}}{6\,081\,075} + \frac{929\,569 t^{15}}{638\,512\,875} + \frac{6\,404\,582 t^{17}}{10\,854\,718\,875} + \frac{443\,861\,162 t^{19}}{1\,856\,156\,927\,625}$

```
In[10]:= Pade[t_] = PadeApproximant[Taylor[t], {t, 0, {10, 10}}]
```

Out[10]= $$\frac{t - \frac{8t^3}{57} + \frac{7t^5}{1615} - \frac{4t^7}{101745} + \frac{t^9}{11904165}}{1 - \frac{9t^2}{19} + \frac{28t^4}{969} - \frac{7t^6}{14535} + \frac{t^8}{440895} - \frac{t^{10}}{654729075}}$$

```
In[11]:= Plot[{Pade[t], Taylor[t], Tan[t]}, {t, 0, 10}, PlotRange → {-5, 5},
  PlotStyle →
   {{AbsoluteThickness[2], AbsoluteDashing[{15, 8}], Black},
    {AbsoluteThickness[4], AbsoluteDashing[{10, 9}], Black},
    {Black}}]
```

Out[11]=

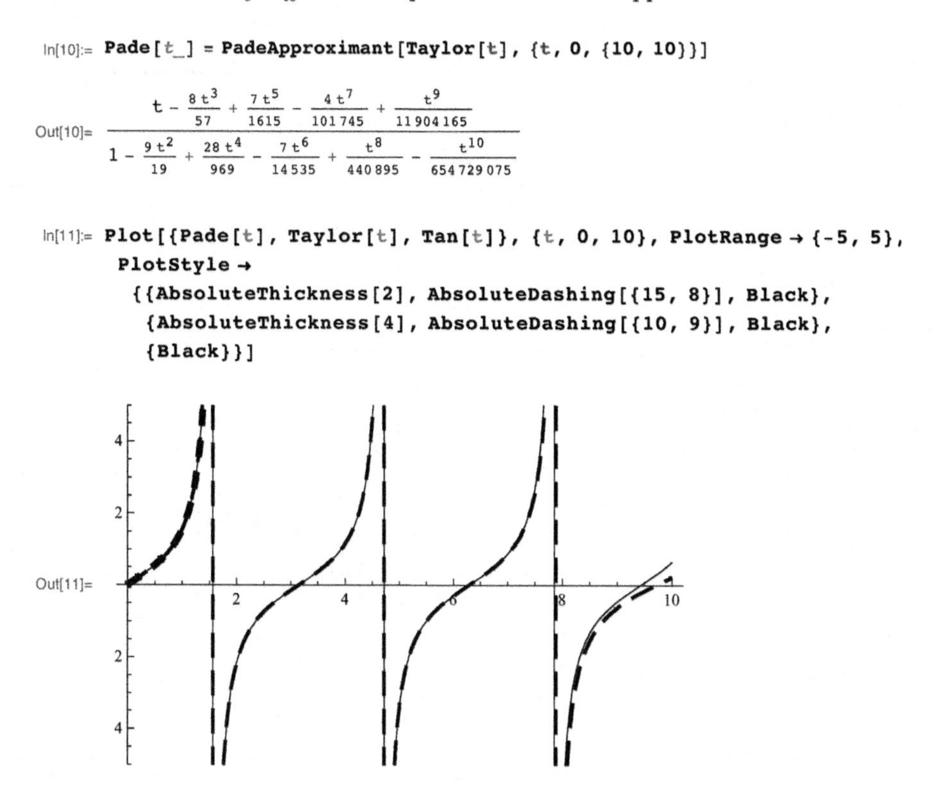

One can see that the Padé approximation does not give satisfactory results for $t > 10$. The Taylor series cannot pass the point $t = \frac{\pi}{2}$ where $\tan x$ has a singularity. It is worth emphasizing that the Taylor polynomial and the Padé approximation are constructed for the same data c_n $(n = 0, 1, \ldots, N)$.

4.6 Harmonic oscillation

4.6.1 Simple harmonic motion

Consider a material point with mass m attached to a massless spring (see Fig.4.5). Let x denote the displacement of the spring from equilibrium. Let the restoring force linearly depend on x, hence $F = -kx$ where k is a positive constant. This equation is called Hooke's law (cf. Example 1.4 on page 20). The potential energy is expressed by (3.14) which takes the form

$$u = \frac{kx^2}{2}. \tag{4.43}$$

Here, the condition $u(0) = 0$ is used. The Law of Conservation of Energy in the phase space (3.17) becomes

$$\frac{mv^2}{2} + \frac{kx^2}{2} = E \Leftrightarrow \frac{m}{2E}v^2 + \frac{k}{2E}x^2 = 1. \quad (4.44)$$

This is an ellipse equation on the plane (x, v). Half-axes of the ellipse are equal to $\sqrt{\frac{2E}{m}}$ and $\sqrt{\frac{2E}{k}}$ (see Fig.4.6).

In the standard course of analytic geometry, the canonical ellipse equation on the plane XOY has the form

$$\frac{x^2}{a^2} + \frac{y^2}{b^2} = 1,$$

where the positive numbers a and b are called half-axes of the ellipse. If $a = b$, we get a circle of the radius a. The ellipse can be written in the parametric form $x = a\cos\theta$, $y = b\sin\theta$, where the parameter θ belongs to the interval $(-\pi, \pi]$.

FIGURE 4.5: A material point and massless spring attached to the celling

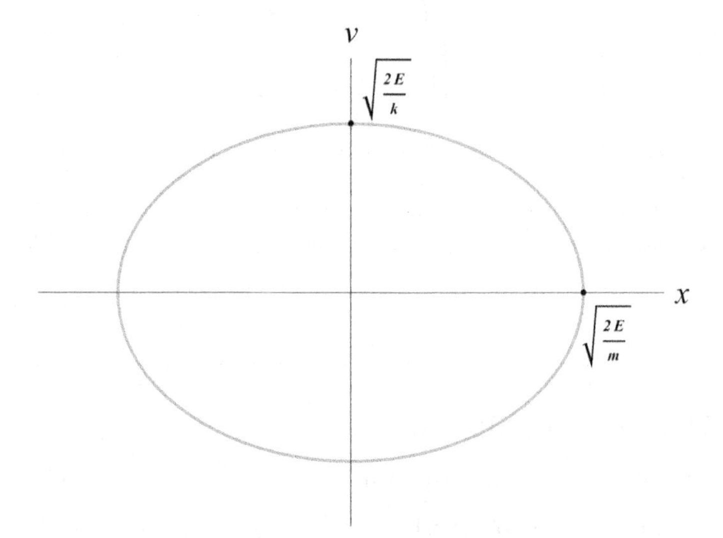

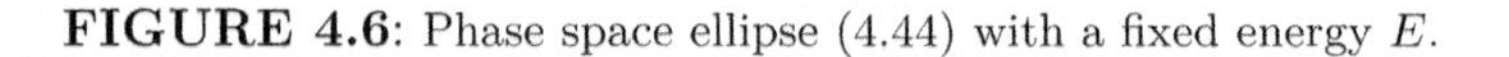

FIGURE 4.6: Phase space ellipse (4.44) with a fixed energy E.

One of the most important observations of Fig.4.6 is that the motion of the particle is periodic. Let the material point be located in a point on the ellipse, i.e., the displacement is equal to x and the velocity to v. Motion of the point in the phase space occurs along an ellipse with a fixed energy E. The point cannot change the energy E without exterior forces and jump over to another ellipse. This yields a periodicity of motion.

In general case, differentiation of (3.16) on the variable x gives (3.20). In the case of equations (4.44), we arrive at the linear ODE with constant coefficients

$$\frac{d^2x}{dt^2} + \omega_0^2 x = 0, \tag{4.45}$$

where the value $\omega_0 = \sqrt{\frac{k}{m}}$ is called *frequency*. Sometimes the frequency is introduced as $\frac{\omega_0}{2\pi}$. Equation (4.45) also follows from (4.44) by the differentiation $\frac{d}{dt}$ and can be directly deduced from the balance of forces acting on the point. The initial conditions are given by formulae

$$x(t_0) = a, \quad x'(t_0) = v_0, \tag{4.46}$$

where v_0 denotes the initial velocity. Without loss of generality we can take $t_0 = 0$. Solve the problem (4.45)–(4.46):

```
In[1]:= DSolve[{x''[t] + ω0^2 x[t] == 0, x[0] == a, x'[0] == v0}, x[t], t]

Out[1]= {{x[t] → (Sin[t ω0] v0 + a Cos[t ω0] ω0)/ω0}}

In[2]:= %[[1, 1, 2]] // Simplify

Out[2]= a Cos[t ω0] + (Sin[t ω0] v0)/ω0
```

The operator **Simplify** simplifies an expression but in a way *Mathematica* wants it, not always corresponding to our wishes. We write the result for Cauchy's problem (4.45)–(4.46)

$$x(t) = a\cos\omega_0 t + \frac{v_0}{\omega_0}\sin\omega_0 t \tag{4.47}$$

in the form

$$x(t) = A\cos(\omega_0 t - \theta_0), \tag{4.48}$$

where the value $A = \sqrt{\frac{v_0^2}{\omega_0^2} + a^2}$ is called *amplitude*, $\theta_0 = \arccos\frac{a}{A}$ the initial phase. The form of the solution (4.48) implies that the motion is periodic with the period $T_0 = \frac{2\pi}{\omega_0}$.

Equation (4.45) was solved with the help of *Mathematica*. But reduction of the trigonometric expression (4.47) to (4.48) was made "by hand". It was done in such a way because it is hard to explain to *Mathematica* what we want and it could take a lot of time. So, the principle of hand calculations from Sec.1.1.3 is used.

The potential energy has the form (see (4.43))

$$u = \frac{kA^2}{2}\cos^2(\omega_0 t - \theta_0). \tag{4.49}$$

The kinetic energy $K = \frac{mv^2}{2} = \frac{m(x'(t))^2}{2}$ becomes

$$K = \frac{kA^2}{2} \sin^2(\omega_0 t - \theta_0). \tag{4.50}$$

The total energy is written as

$$E = u + K = \frac{kA^2}{2} = \frac{mv_0^2}{2} + \frac{ka^2}{2}. \tag{4.51}$$

4.6.2 Harmonic oscillator with friction and exterior forces

In the present subsection, the general scheme from Sec.1.1.3 is applied to modify the model (4.45) by introduction of friction and exterior forces. It follows from experiments that the friction force is proportional to the velocity of the material point, i.e., $F_{friction} = -hv$ where h is a positive constant. Introduction of $F_{friction}$ into (4.45) yields the linear differential equation with constant coefficients

$$\frac{d^2x}{dt^2} + 2\gamma\frac{dx}{dt} + \omega_0^2 x = 0, \tag{4.52}$$

where another constant $2\gamma = \frac{h}{m}$ is introduced. Equation (4.52) models *damped motion* with friction.

The general solution of (4.52) depends on two arbitrary constants `C[1]` and `C[2]`:

```
In[1]:= DSolve[x''[t] + 2 γ x'[t] + ω0^2 x[t] == 0, x[t], t]

Out[1]= {{x[t] → e^(t (-γ - Sqrt[γ^2 - ω0^2])) C[1] + e^(t (-γ + Sqrt[γ^2 - ω0^2])) C[2]}}
```

Let an exterior force F_{ext} act on the particle. Then, the right-hand side of equation (4.52) has to contain this force:

$$\frac{d^2x}{dt^2} + 2\gamma\frac{dx}{dt} + \omega_0^2 x = \frac{F_{ext}}{m}. \tag{4.53}$$

As an example, consider the periodic in time force $\frac{F_{ext}}{m} = f\cos(\omega t)$ where f is a constant, ω is a frequency of the given exterior force. Then, we arrive at the nonhomogeneous linear differential equation with constant coefficients

$$\frac{d^2x}{dt^2} + 2\gamma\frac{dx}{dt} + \omega_0^2 x = f\cos(\omega t). \tag{4.54}$$

```
In[2]:= DSolve[x''[t] + 2 γ x'[t] + ω0^2 x[t] == f Cos[ω t], x[t], t]
```

$$\text{Out[2]= } \left\{\left\{x[t] \to e^{t\left(-\gamma-\sqrt{\gamma^2-\omega_0^2}\right)} C[1] + e^{t\left(-\gamma+\sqrt{\gamma^2-\omega_0^2}\right)} C[2] + \frac{f\,\omega^2 \text{Cos}[t\,\omega] - 2 f\,\gamma\,\omega\,\text{Sin}[t\,\omega] - f\,\text{Cos}[t\,\omega]\,\omega_0^2}{\left(2\gamma^2+\omega^2-\omega_0^2+2\gamma\sqrt{\gamma^2-\omega_0^2}\right)\left(-2\gamma^2-\omega^2+\omega_0^2+2\gamma\sqrt{\gamma^2-\omega_0^2}\right)}\right\}\right\}$$

```
In[3]:= %[[1, 1, 2]] // Simplify
```

$$\text{Out[3]= } e^{-t\left(\gamma+\sqrt{\gamma^2-\omega_0^2}\right)} C[1] + e^{t\left(-\gamma+\sqrt{\gamma^2-\omega_0^2}\right)} C[2] + \frac{f\left(\omega\left(-\omega\,\text{Cos}[t\,\omega] + 2\gamma\,\text{Sin}[t\,\omega]\right) + \text{Cos}[t\,\omega]\,\omega_0^2\right)}{4\gamma^2\omega^2+\omega^4-2\omega^2\omega_0^2+\omega_0^4}$$

The latter formula gives the general solution of equation (4.54) with arbitrary constants C[1] and C[2]. Let the initial conditions for equation (4.54) have the form $x(0) = 1$, $x'(0) = 0$. Then, the operators In[2] - In[3] are corrected as follows

```
In[4]:= Clear[f]; X0[t_, ω_, ω0_, γ_, f_] =
          DSolve[{x''[t] + 2 γ x'[t] + ω0^2 x[t] == f Cos[ω t], x[0] == 1,
              x'[0] == 0}, x[t], t][[1, 1, 2]] // Simplify
```

$$\text{Out[4]= } \Big(e^{-t\left(\gamma+\sqrt{\gamma^2-\omega0^2}\right)} \Big(f\gamma\omega^2 - e^{2t\sqrt{\gamma^2-\omega0^2}} f\gamma\omega^2 - 4\gamma^3\omega^2 +$$
$$4 e^{2t\sqrt{\gamma^2-\omega0^2}}\gamma^3\omega^2 - \gamma\omega^4 + e^{2t\sqrt{\gamma^2-\omega0^2}}\gamma\omega^4 + f\gamma\omega0^2 -$$
$$e^{2t\sqrt{\gamma^2-\omega0^2}} f\gamma\omega0^2 + 2\gamma\omega^2\omega0^2 - 2e^{2t\sqrt{\gamma^2-\omega0^2}}\gamma\omega^2\omega0^2 - \gamma\omega0^4 +$$
$$e^{2t\sqrt{\gamma^2-\omega0^2}}\gamma\omega0^4 + f\omega^2\sqrt{\gamma^2-\omega0^2} + e^{2t\sqrt{\gamma^2-\omega0^2}} f\omega^2\sqrt{\gamma^2-\omega0^2} +$$
$$4\gamma^2\omega^2\sqrt{\gamma^2-\omega0^2} + 4e^{2t\sqrt{\gamma^2-\omega0^2}}\gamma^2\omega^2\sqrt{\gamma^2-\omega0^2} +$$
$$\omega^4\sqrt{\gamma^2-\omega0^2} + e^{2t\sqrt{\gamma^2-\omega0^2}}\omega^4\sqrt{\gamma^2-\omega0^2} -$$
$$f\omega0^2\sqrt{\gamma^2-\omega0^2} - e^{2t\sqrt{\gamma^2-\omega0^2}} f\omega0^2\sqrt{\gamma^2-\omega0^2} -$$
$$2\omega^2\omega0^2\sqrt{\gamma^2-\omega0^2} - 2e^{2t\sqrt{\gamma^2-\omega0^2}}\omega^2\omega0^2\sqrt{\gamma^2-\omega0^2} +$$
$$\omega0^4\sqrt{\gamma^2-\omega0^2} + e^{2t\sqrt{\gamma^2-\omega0^2}}\omega0^4\sqrt{\gamma^2-\omega0^2} +$$
$$2e^{t\left(\gamma+\sqrt{\gamma^2-\omega0^2}\right)} f\sqrt{\gamma^2-\omega0^2}\left(-\omega^2+\omega0^2\right)\text{Cos}[t\,\omega] +$$
$$4e^{t\left(\gamma+\sqrt{\gamma^2-\omega0^2}\right)} f\gamma\omega\sqrt{\gamma^2-\omega0^2}\,\text{Sin}[t\,\omega]\Big)\Big)\Big/$$
$$\left(2\sqrt{\gamma^2-\omega0^2}\left(4\gamma^2\omega^2+\left(\omega^2-\omega0^2\right)^2\right)\right)$$

Out[5]=

$$\Big(e^{-t\left(\gamma+\sqrt{\gamma^2-\omega 0^2}\right)}\Big(-2\,f\,\gamma^2\,\omega^2+2\,e^{2\,t\,\sqrt{\gamma^2-\omega 0^2}}\,f\,\gamma^2\,\omega^2+f\,\omega^2\,\omega 0^2-e^{2\,t\,\sqrt{\gamma^2-\omega 0^2}}\,f\,\omega^2\,\omega 0^2+$$
$$4\,\gamma^2\,\omega^2\,\omega 0^2-4\,e^{2\,t\,\sqrt{\gamma^2-\omega 0^2}}\,\gamma^2\,\omega^2\,\omega 0^2+\omega^4\,\omega 0^2-e^{2\,t\,\sqrt{\gamma^2-\omega 0^2}}\,\omega^4\,\omega 0^2-$$
$$f\,\omega 0^4+e^{2\,t\,\sqrt{\gamma^2-\omega 0^2}}\,f\,\omega 0^4-2\,\omega^2\,\omega 0^4+2\,e^{2\,t\,\sqrt{\gamma^2-\omega 0^2}}\,\omega^2\,\omega 0^4+$$
$$\omega 0^6-e^{2\,t\,\sqrt{\gamma^2-\omega 0^2}}\,\omega 0^6-2\,f\,\gamma\,\omega^2\,\sqrt{\gamma^2-\omega 0^2}-2\,e^{2\,t\,\sqrt{\gamma^2-\omega 0^2}}\,f\,\gamma$$
$$\omega^2\,\sqrt{\gamma^2-\omega 0^2}+4\,e^{t\left(\gamma+\sqrt{\gamma^2-\omega 0^2}\right)}\,f\,\gamma\,\omega^2\,\sqrt{\gamma^2-\omega 0^2}\,\mathrm{Cos}[t\,\omega]-$$
$$2\,e^{t\left(\gamma+\sqrt{\gamma^2-\omega 0^2}\right)}\,f\,\omega\,\sqrt{\gamma^2-\omega 0^2}\,\left(-\omega^2+\omega 0^2\right)\,\mathrm{Sin}[t\,\omega]\Big)\Big)\Big/$$
$$\left(2\,\sqrt{\gamma^2-\omega 0^2}\,\left(4\,\gamma^2\,\omega^2+\left(\omega^2-\omega 0^2\right)^2\right)\right)$$

In[6]:=

```
ParametricPlot[{X0[t, 1.1, 1, 0.1, 1], X1[t, 1.1, 1, 0.1, 1]},
  {t, 1, 50}, ImageSize → 400, PlotStyle → Black, AxesStyle → 12,
  AspectRatio → Automatic]
```

Out[6]=

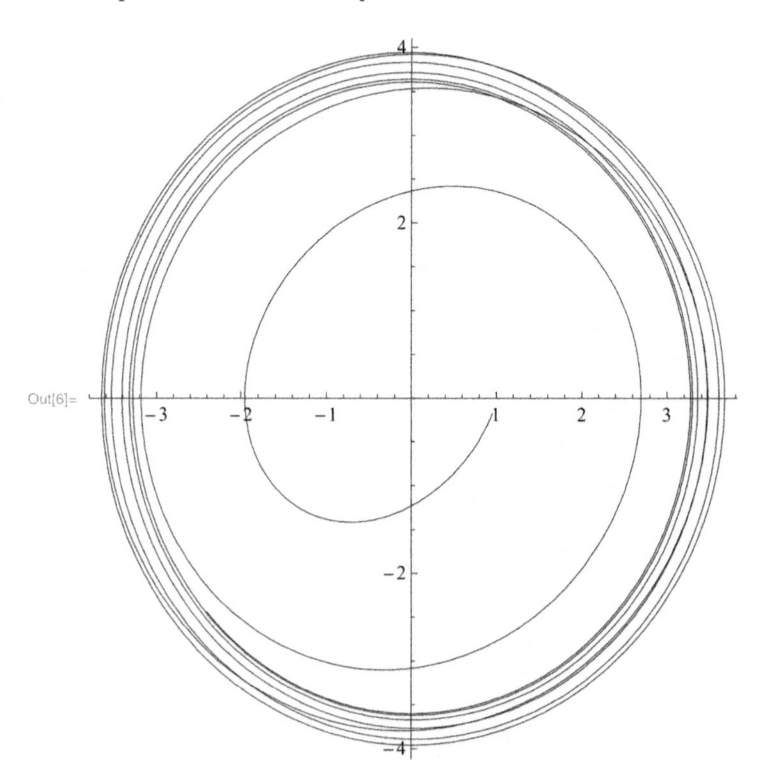

The operator **ParametricPlot** draws a curve on the phase plane (x, v) determined by the parametric equations $x = x(t)$, $v = v(t)$.

MATLAB Example Box 4.2

One can obtain similar results using MATLAB script. Let us also present how to change the accuracy of `ode45` operator via the `odeset` function. In the example below, the relative error tolerance (`RelTol`) is set to 10^{-2} (the default value is 10^{-3}). Observe how the solution changes by setting different values, say `1e-4`, `1e-1`.

```
function script11()

    % parameters
    omega = 1.1;
    omega0 = 1;
    gamma = 0.1;
    f = 1;
    t_interval = [0 20];

    % solution of the equation
    g = @(t, x) g(t, x, gamma, omega, omega0, f);
    options = odeset('RelTol',1e-2, 'AbsTol', 1e-6);
    [T, X] = ode45(g, t_interval, [1 0], options);

    % parametric plot (x1(t), x2(t))
    plot(X(:,1), X(:,2), '-o')
    xlabel('x1')
    ylabel('x2')
    axis equal
    grid on

function dx = g(t, x, gamma, omega, omega0, f)
    dx = zeros(2,1);
    dx(1) = x(2);
    dx(2) = -2*gamma * x(2) - omega0^2 * x(1) ...
            + f * cos(omega * t);
```

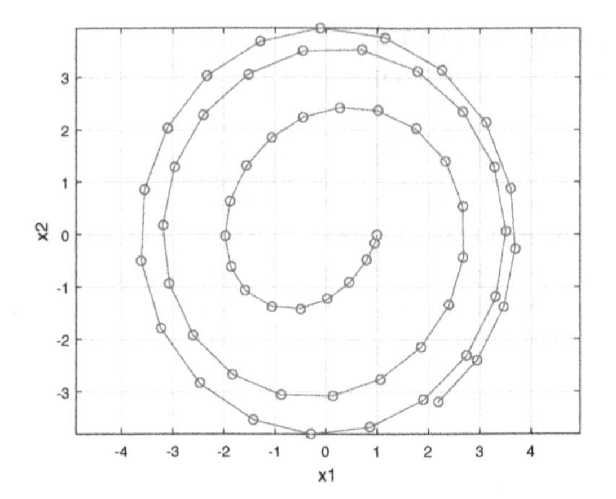

The `RelTol` and `AbsTol` are examples of options responsible for error control. For more information on how these options work, see the documentation on `odeset` function.

It is worth noting that symbolic computations can yield too long formulae much longer than the ones above. This makes them useless for traditional applications since we observe a short formula trying to note its features. Moreover, visualization of long formulae can be very expensive because of the memory occupied by symbols. Fortunately, we need not see a formula in order to work with it. This is an important new point in symbolic computations. For instance, the inputs In[4] and In[5] on page 77 can be ended by the symbol "`;`". Then, the expressions **X0** and **X1** will be introduced into the kernel of *Mathematica* but will be not shown. Further, the input In[6] will display the same picture. Hence, one can manipulate **X0** and **X1** without their observation. It is possible to look at a fragment of **X0**. For instance, the operator **Coefficient[X,ω²]** selects all the coefficients on ω^2 in the expression **X**.

4.6.3 Resonance

Consider a harmonic oscillator without friction but with the exterior force $f\cos\omega t$ and zero initial conditions. We have

```
In[7]:= X2[t_, ω_, ω0_, f_] =
          DSolve[{x''[t] + ω0^2 x[t] == f Cos[ω t], x[0] == 1, x'[0] == 0},
            x[t], t][[1, 1, 2]] // Simplify
```

$$\text{Out[7]=}\quad \frac{f\,\text{Cos}[t\,\omega] - (f + \omega^2 - \omega0^2)\,\text{Cos}[t\,\omega0]}{-\omega^2 + \omega0^2}$$

The denominator of the solution Out[7] can be equal to zero.

Let us assume that

```
In[8]:= ω = ω0

Out[8]= ω0
```

The latter equality means that the frequency ω of the exterior force coincides with *the natural frequency* ω_0. This leads to a phenomenon called *resonance.* In this case, the solution **X2[t,ω0,ω0,f]** fails. We have equation

$$\frac{d^2x}{dt^2} + \omega_0^2 x = f\cos(\omega_0 t) \tag{4.55}$$

and its solution

```
In[9]:= X3[t_, ω0_, f_] =
          DSolve[{x''[t] + ω0² x[t] == f Cos[ω t], x[0] == 1, x'[0] == 0},
            x[t], t][[1, 1, 2]] // Simplify

                           f t Sin[t ω0]
Out[9]= Cos[t ω0] + ─────────────
                               2 ω0
```

The amplitude of the latter solution has the form

```
In[10]:= A3[t_, ω0_, f_] = √(1 + (f t / (2 ω0))²) ;
```

One can see that the amplitude in the resonance case increases indefinitely with the time contrary to the case $\omega \neq \omega_0$.

Consider a more complicated equation which takes into account friction with a coefficient γ

$$\frac{d^2x}{dt^2} + 2\gamma\frac{dx}{dt} + \omega^2 x = f\cos(\omega_0 t). \tag{4.56}$$

The resonance phenomenon can be observed by investigation of the solution of (4.56) near the point $\omega = \omega 0$. Put $f = 1$. The amplitude of the solution has the form

```
In[11]:= Clear[ω]; f = 1;
         A[ω_, ω0_, γ_] =
          √(Coefficient[X0[t, ω, ω0, γ, f], Cos[t ω]]² +
              Coefficient[X0[t, ω, ω0, γ, f], Sin[t ω]]²) // Simplify

Out[11]= √(1 / (4 γ² ω² + (ω² - ω0²)²))

In[12]:= Plot[{A[ω, 1, 0.05], A[ω, 1, 0.1], A[ω, 1, 0.2]}, {ω, 0.8, 1.2},
          PlotStyle → {{Thick, Gray}, {Thick, Dashed, Black},
            {Thick, Dotted, Black}}, PlotRange → All, AxesLabel → {ω, A},
          LabelStyle → 16, TicksStyle → 12]
```

Out[12]=
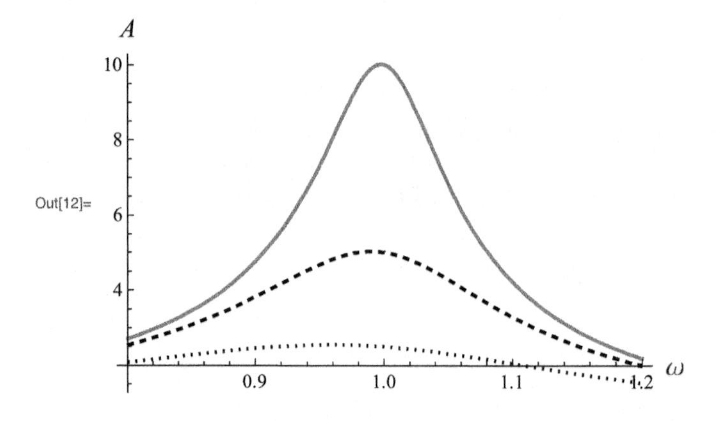

The maximal value of the amplitude is found exactly by the standard calculus investigation of the function $4\gamma^2\omega^2 + (\omega^2 - \omega_0^2)^2$ to determine its minimum on the variable ω. Calculate the derivative and equal it to zero:

```
In[13]:= Dm[ω_, ω0_, γ_] = ∂ω (4 γ² ω² + (ω² - ω0²)²)

Out[13]= 8 γ² ω + 4 ω (ω² - ω0²)

In[14]:= Solve[Dm[ω, ω0, γ] == 0, ω]

Out[14]= {{ω → 0}, {ω → -√(-2 γ² + ω0²)}, {ω → √(-2 γ² + ω0²)}}
```

Only the third root $\omega = \sqrt{-2\gamma^2 + \omega_0^2}$ satisfies the inequality $2\gamma^2 < \omega_0^2$.

Modify the graph by addition of the maximal point

```
In[15]:= Mpoint[ω0_, γ_] = {√(-2 γ² + ω0²), A[√(-2 γ² + ω0²), ω0, γ]}

Out[15]= {√(-2 γ² + ω0²), √(1/(4 γ⁴ + 4 γ² (-2 γ² + ω0²)))}

In[16]:= Plot[{A[ω, 1, 0.05], A[ω, 1, 0.1], A[ω, 1, 0.2]}, {ω, 0.7, 1.3},
  PlotStyle → {{Thick, Gray}, {Thick, Dashed, Black},
    {Thick, Dotted, Black}}, PlotRange → {0, 11}, AxesLabel → {ω, A},
  Epilog → {{PointSize[0.02], Point[Mpoint[1, 0.05]]},
    {PointSize[0.02], Point[Mpoint[1, 0.1]]},
    {PointSize[0.02], Point[Mpoint[1, 0.2]]}}, GridLines → Automatic,
  LabelStyle → 16, TicksStyle → 12, GridLinesStyle → Gray]
```

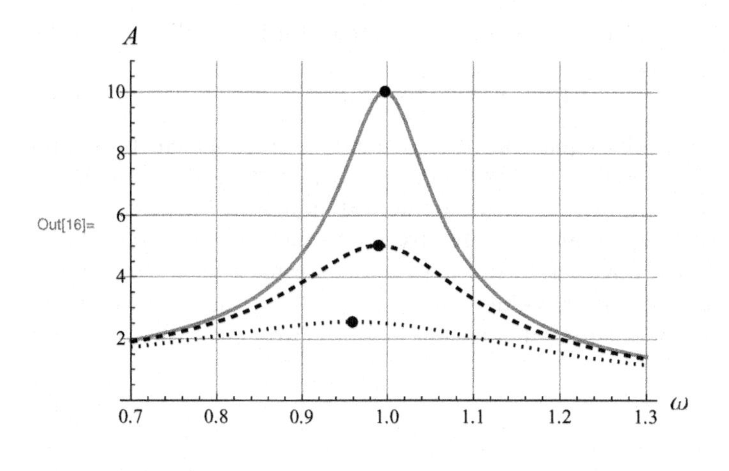

4.7 Lotka-Volterra model

The Lotka-Volterra model describes predator-prey systems when one biological family chases another family (wolves chase rabbits, pikes chase crucians). Let the number of pikes be denoted by $y(t)$ and the number of crucians by $x(t)$. If we consider only one independent group of crucians, it can be described by the differential equation (4.15), i.e., by equation

$$x'(t) = kx(t). \tag{4.57}$$

Introduction of pikes into the pond yields modifications of the model (4.57) because of the eaten crucians. The number of fatal meetings is proportional to $x(t)$ and to $y(t)$. Then, equation (4.57) transforms to

$$\frac{dx}{dt} = kx - axy, \tag{4.58}$$

where a is a positive constant. Pikes die out without food that can be modeled by equation

$$y'(t) = -\ell y(t), \tag{4.59}$$

where ℓ is a positive constant. The presence of crucians gives an additional term to this equation proportional to the number of fortunate meetings:

$$\frac{dy}{dt} = -\ell y + bxy, \tag{4.60}$$

where b is a positive constant.

The system of ordinary differential equations (4.58), (4.60) is called the

Lotka-Volterra model of the predator-prey system. Initial conditions can be added to equations

$$x(0) = x_0, \quad y(0) = y_0. \tag{4.61}$$

It is impossible to solve equations (4.58), (4.60) in terms of elementary and special functions and possible to solve them numerically. However, many interesting features will be hidden in such a numerical approach. We now proceed to apply the phase space to analytically investigate the system (4.58), (4.60). The phase space is the plane (x, y) in our case that formally diverges with the definition given in Sec.3.4[2]. Dividing equation (4.58) by equation (4.60) we arrive at the ordinary differential equation which relates two variables x and y

$$\frac{dy}{dx} = \frac{y(-\ell + bx)}{x(k - ay)}. \tag{4.62}$$

Time has been disappeared due to the formal independence of the right-hand sides of (4.58), (4.60) on t. Equation (4.62) is solved by separation of variables (see Sec.8.5). We have

$$\frac{k - ay}{y}\, dy = \frac{-\ell + bx}{x}\, dx. \tag{4.63}$$

Integration of (4.63) gives the general solution of (4.63) in the form of the implicit function

$$ay - k \ln y + bx - \ell \ln x = C, \tag{4.64}$$

where C is an arbitrary constant. The variables x and y are non negative since they express the numbers of beings.

The function $f(x) = bx - \ell \ln x$ in the single variable x can be investigated by the standard method of calculus. Its derivative $f'(x) = b - \frac{\ell}{x}$ has one critical point $x = \frac{\ell}{b}$. Moreover, $f'(x) < 0$, hence $f(x)$ decreases for $0 < x < \frac{\ell}{b}$ and $f'(x) > 0$, hence $f(x)$ increases for $\frac{\ell}{b} < x < \infty$. Therefore, $f(x)$ attains the global minimum at $x = \frac{\ell}{b}$. The function $g(y) = ay - k \ln y$ has analogous properties. The critical point holds $y = \frac{k}{a}$. The global minimum of the function $f(x) + g(y)$ is attained at the point $Q = \left(\frac{\ell}{b}, \frac{k}{a}\right)$. It is equal to $C_0 = k(1 - \ln \frac{k}{a}) + \ell(1 - \frac{\ell}{b})$. Consider equation (4.64) as an equation of the surface defined by the function $C = C(x, y)$. Then, the curve (4.64) can be displayed on the plane XOY through the section $C = constant$ of the surface $C = C(x, y)$. One can see that the equality $C = C_0$ is possible only at the point Q and the values $C > C_0$ determine closed ovals around Q since the function $f(x) + g(y)$ has only one local minimum at Q.

The typical curves (4.64) are displayed in Fig.4.7. One can easily describe the cyclic character of the system by taking an initial condition (4.61) in the following way. Mark an initial point A that fixes a curve on Fig.4.7. The point

[2] We explain that the phase space (x, y) is consistent with the definition at the end of the section.

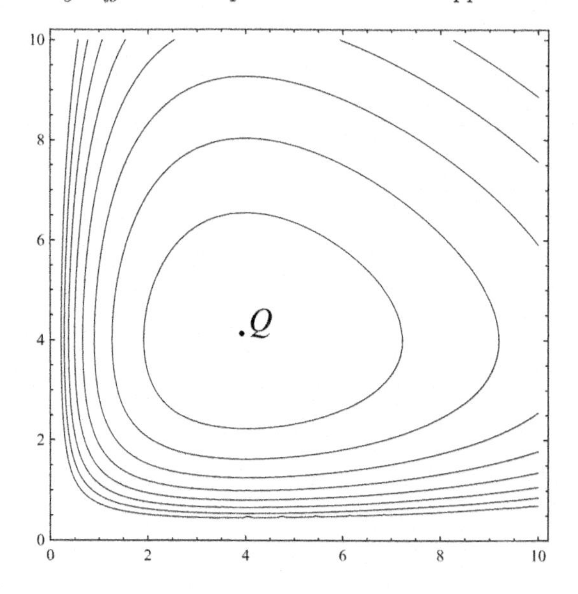

FIGURE 4.7: The phase portrait of equations (4.58), (4.60) for $a = 1.5$, $k = 6$, $b = 1$, $\ell = 4$. The stationary point $Q = (4, 4)$.

A goes on the curve with time t and comes back at the initial position. It is useful to analyze the values x and y going along the curve. Biologists made various corollaries from their observations before the creation of the Lotka-Volterra model. For instance, observation of the wolf populations showed their "strange" oscillations, but only the Lotka-Volterra model allowed one to make correct explanations of the observed results.

It should be added that the Lotka-Volterra model also describes chemical reactions of two ingredients of concentrations x and y. This approach develops in the framework of chemical physics and engineering [41], [22].

Remark 4.2. Actually, there is no any contradiction between the definition of the phase space (x, v) given in Sec.3.4 and (x, y) in this section. Formally in this section, we have to consider the 4-dimensional phase space (x, y, x', y') where $x' = \frac{dx}{dt}$ and $y' = \frac{dy}{dt}$. This space is decomposed into the plane portrait (x, y) and the dependence (x', y') on (x, y) expressed through equations (4.58), (4.60). The part (x, y) tells us sufficiently much. The part (x', y') refines the general picture. For instance, we can determine in which direction point A goes on the oval and estimate its velocity.

MATLAB Example Box 4.3

Let us plot a level phase portrait of the Lotka-Volterra model with `contour` and `meshgrid` operators.

```
function script12()
```

```
% parameters
a = 1.5;
k = 6;
b = 1;
l = 4;

% 2D grid coordinates
x = linspace(0, 10);
y = linspace(0, 10);
[X,Y] = meshgrid(x,y);
% height values corresponding to the grid points
C = a*Y -k*log(Y) + b*X -l*log(X);

cplot = contour(X,Y,C,20);
```

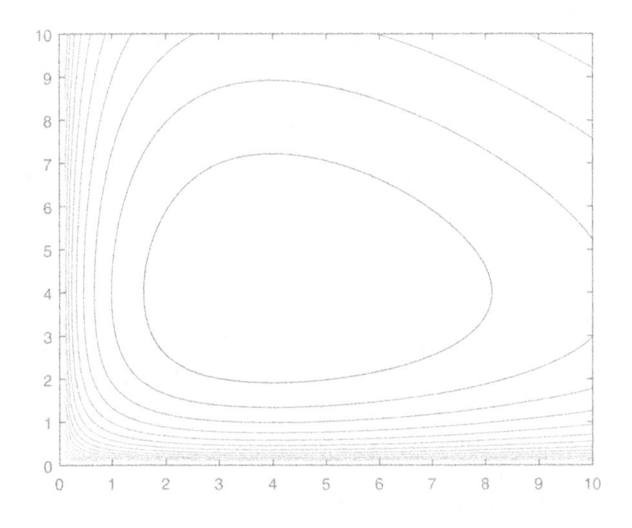

Here, `meshgrid` operator generates 2D grid coordinates in the form of two matrices `X` and `Y` containing copies of vectors `x` and `y`, respectively (examine the output of `meshgrid` by yourself). Then, the `contour` plots 20 contour levels based on the `C` matrix incorporating values of height corresponding to coordinates in the XOY-plane.

4.8 Linearization

The method of linearization was applied in Sec.2.3 to Newton's method when the nonlinear equation (2.5) was approximated by the linear one (2.6). The method of linearization is a general constructive method of mathematical modeling. It can be applied to differential equations as is done in Sec.4.3 .

In this section, we discuss the nonlinear system of ODE

$$\frac{dx}{dt} = f(x, y), \quad \frac{dy}{dt} = g(x, y), \tag{4.65}$$

where the unknown functions $x = x(t)$, $y = y(t)$ are differentiable in the segment $(0, T)$ and the functions $f(x, y)$, $g(x, y)$ are given. Let (x_0, y_0) be a fixed *equilibrium point* of the system (4.65), i.e.,

$$f(x_0, y_0) = 0, \quad g(x_0, y_0) = 0. \tag{4.66}$$

It is assumed that $f(x, y)$ and $g(x, y)$ are continuously differentiable in a vicinity of (x_0, y_0). Introduce new unknowns $u(t) = x(t)$ and $v(t) = y(t)$. Then, equations (4.65) become

$$\frac{du}{dt} = F(u, v), \quad \frac{dv}{dt} = G(u, v), \tag{4.67}$$

where $F(u, v) = f(x, y)$ and $G(u, v) = g(x, y)$. Using the approximations for sufficiently small u and v

$$F(u, v) \approx \frac{\partial F}{\partial u}u + \frac{\partial F}{\partial v}v \text{ and } G(u, v) \approx \frac{\partial G}{\partial u}u + \frac{\partial G}{\partial v}v \tag{4.68}$$

we arrive at the linear differential equations

$$\frac{du}{dt} = a_{11}u + a_{12}v, \quad \frac{dv}{dt} = a_{21}u + a_{22}v, \tag{4.69}$$

where

$$A = \begin{pmatrix} a_{11} & a_{12} \\ a_{21} & a_{22} \end{pmatrix} = \begin{pmatrix} \frac{\partial F}{\partial u} & \frac{\partial F}{\partial v} \\ \frac{\partial G}{\partial u} & \frac{\partial G}{\partial v} \end{pmatrix}. \tag{4.70}$$

The partial derivatives in (4.70) are calculated at the point (x_0, y_0), hence the matrix A does not depend on t. The matrix (4.70) is called the *Jacobi matrix*. It is supposed that A is not degenerate, i.e., $\det A \neq 0$. Introduce the vector-function $\mathbf{u} = (u, v)^T$ and write (4.69) in the vector-matrix form

$$\frac{d\mathbf{u}}{dt} = A\mathbf{u}. \tag{4.71}$$

The theory of linear ODE with constant coefficients suggests to look for the vector-function $\mathbf{u}$ in the form

$$\mathbf{u}(t) = \mathbf{v}e^{\lambda t}, \tag{4.72}$$

where the constant vector $\mathbf{v}$ and number λ have to be determined. Substitution of (4.72) into (4.71) yields the *eigenvalue problem*

$$A\mathbf{v} = \lambda\mathbf{v}. \tag{4.73}$$

We have to find a constant λ called the eigenvalue and vectors $\mathbf{v}$ called the eigenvectors associated to the eigenvalue λ. According to the standard course of linear algebra the eigenvalues satisfy an algebraic equation, in our two-dimensional case, to the quadratic equation

$$\det(A - \lambda I) = 0 \quad \Longleftrightarrow \quad \lambda^2 - (a_{11} + a_{22})\lambda - a_{12}a_{21} = 0, \tag{4.74}$$

where I denotes the identity matrix. When the roots λ_1 and λ_2 of (4.74) are found, the eigenvector $\mathbf{v}_1 = (v_1, v_2)$ corresponding to λ_1 can be determined from equation

$$(A - \lambda_1 I)\mathbf{v}_1 = \mathbf{0} \quad \Longleftrightarrow \quad a_{21}v_1 + (a_{22} - \lambda)v_2 = 0, \tag{4.75}$$

Putting $v_1 = 1$ we get $v_2 = \frac{a_{21}}{\lambda_1 - a_{22}}$, hence $\mathbf{v}_1 = \left(1, \frac{a_{21}}{\lambda_1 - a_{22}}\right)$. Further study depends on the roots λ_1 and λ_2 of equation (4.74). It is worth noting that the case $a_{12} = a_{21}$ (the matrix A is symmetric) frequently occurs in practice. Then, the eigenvalues are always real and different.

Let $0 < \lambda_1 < \lambda_2$. Then, the general solution of (4.74) is given by formula

$$\mathbf{u}(t) = C_1\mathbf{v}_1 e^{\lambda_1 t} + C_2\mathbf{v}_2 e^{\lambda_2 t}, \tag{4.76}$$

where C_1 and C_2 are arbitrary constants. The vector-function (4.76) is not bounded for any C_1 and C_2 since λ_1 and λ_2 are positive. Therefore, in this case the solution of the linearized equation is not stable in time and the equilibrium point (x_0, y_0) is repulsive for the original equations (4.65).

Other cases are investigated by similar arguments. As a result of such a complete study we get the following necessary and sufficient stability condition valid for arbitrary dimension of A. The system (4.71) is stable if and only if the eigenvalues of A have non-positive real parts.

Concluding remarks and further reading. This chapter demonstrates that ODE can be applied to various fields of human activity up to human emotions and feelings as in Exercise 17 below. We refer [5, 22, 32, 33, 48, 49, 52, 55, 61] for further reading. This list is not complete.

Exercises

1. Prepare a project with animation for Example 4.1.

2. Invite a special function or a class of special functions by the linguistic method following Bessel and his equation (4.1).

3. A reservoir contains 10 liter of air (20% of oxygen and 80% of nitrogen).

Let 0.2 litre of oxygen enter per minute into the reservoir and 0.2 litre of mixture go out. When has the oxygen concentration achieved the concentration 99.9%?

4. Let a kettle be cooled down from 90° (in Celsius) to 50° during 5 min. in the room with the temperature 20°. When will the kettle cool down up to 20°?

5. Present a graphs of (4.20) in one figure and describe types of the logistic curves for different parameters x_0, a and k using the operator **Manipulate**.

   As an advanced exercise, you can create slider manipulator in MATLAB. Refer to the documentation and investigate the `uicontrol` operator.

6. Calculate numerically the sums $\sum_{n=1}^{\infty} \frac{1}{n^m}$ for $m = 1, 2, 3, 4, \ldots$.

7. Perform dimensional analysis of equations (4.21) and (4.23).

8. Describe the set of parameters r and q for which a root of equation (4.25) is positive. Investigate all the roots.

9. Investigate the model (4.21) when the function (4.22) is replaced by $p(x) = \frac{\alpha x}{\beta + x}$ or by $p(x) = \alpha \arctan \beta x$.

10. Which k and t_0 equation (4.35) has only real roots? Investigate the number of complex roots of equation (4.35) and investigate their locations.

    Hint: In order to reduce the number of parameters, first make the change $\lambda = k\nu$.

    The number of complex roots can be infinite. They accumulate to infinity.

    Further analysis of the delay model can be made following Murray [48, Section 1.3 and 1.4].

11. Cauchy's problem (4.37) for $m = 3$ can be numerically solved by the operator **NDSolve** for $t \in (0, T)$. Determine the maximally possible value T for which **NDSolve** gives an acceptable solution. Compare the results with the approximations obtained in Sec.4.5.1-4.5.2.

    Try to solve the equation in MATLAB with `ode45` operator. Refer to the Example Box 4.1 on page 63.

12. Solve Cauchy's problem (4.37) for $m = 4$, 5, 0.5 and 1.5.

13. Present the harmonic oscillator as a hanging mass on a spring attached to the ceiling. Prepare two parallel graphical animations demonstrating the vertical oscillations as in Fig.4.5 and the corresponding graph of the motion of the mass point in time.

    For a hint, see demonstration projects at `http://demonstrations.wolfram.com/`

14. Investigate the harmonic oscillator with fraction and exterior forces determined by the functions $f\sin(\omega t)$ and $f\exp(kt)$ where k is a real parameter. Investigate the corresponding resonance problem.

15. Equations (4.52) and (4.54) have been solved for $\gamma > \omega_0$ (see the expression $\sqrt{\gamma^2 - \omega_0^2}$ in the corresponding outputs). Solve these equations for $\gamma \leq \omega_0$ and compare the solutions.

16. Solve equations $x''(t) + 4x(t) = \cos 1.95t$ and $x''(t) + 4x(t) = \cos 2t$ with the conditions $x(0) = 0$ and $x'(0) = 0$. Compare their solutions in the intervals $(0, \pi)$ and $(0, 100\pi)$.

17. Develop a model of love. Let $x(t)$ denote the love intensity of Romeo to Juliette and $y(t)$ of Juliette to Romeo. Let love changes with $x(t)$ and $y(t)$, for instance, $x'(t)$ is a linear combination of $x(t)$ and $y(t)$. Using movies, literature and own life experience fit coefficients for the differential equations of love. Construct the corresponding graphs.

    Modify the model by addition of the third side, for instance Othello or Paris with the corresponding coefficients.

    Investigate the question, for which number of participants the model becomes chaotic.

18. Consider the fishing problem for the Lotka-Volterra model [32]. Let fishing extract the same fraction δ per number of pikes and crucians. Then, the coefficients k and ℓ in (4.58) and (4.60) become $k + \delta$ and $\ell + \delta$. Investigate the dependence of the system on δ.

    Let T denote a time period over a phase cycle curve. Introduce the averaged populations of pikes and crucians over a cycle

$$\overline{x} = \frac{1}{T}\int_0^T x(t)\,\mathrm{d}t, \quad \overline{y} = \frac{1}{T}\int_0^T y(t)\,\mathrm{d}t. \tag{4.77}$$

    Prove that $\overline{x} = \frac{\ell}{b}, \quad \overline{y} = \frac{k}{a}$.

    Hint: Find $\frac{x'(t)}{x(t)}$ from (4.58), calculate the integral $\frac{1}{T}\int_0^T \frac{x'(t)}{x(t)}\,\mathrm{d}t$ and use the periodicity of $x(t)$ on the interval $(0, T)$.

    Estimate the change of the averaged populations of pikes and crucians on the fishing fraction δ.

    Prepare a demonstration including graphic animations of $x(t)$, $y(t)$, $x'(t)$, $y'(t)$ and of the phase portrait of the system.

19. C.S. Holling modified the Lotka-Volterra model by restriction of the infinite voracity of predators expressed by the terms axy and bxy in (4.58), (4.60). The modified equations have the form

$$\frac{dx}{dt} = rx\left(1 - \frac{x}{k}\right) - \frac{\omega xy}{d + x}, \quad \frac{dy}{dt} = sy\left(1 - \frac{Jy}{x}\right). \tag{4.78}$$

Analyse Holling's model following Sec.4.7.

20. Develop a discrete linear model of the capitalization in a bank system.

    Hint: Let x_t denote a sum of money at time t measured in years in a bank. Then, $x_{t+1} = (1 + p)x_t$ where p denotes the annual rate.

Chapter 5

Stochastic models

5.1 Method of least squares 93
5.2 Fitting 97
5.3 Method of Monte Carlo 102
5.4 Random walk 105
Exercises 109

5.1 Method of least squares

We now proceed with discussion of a method how to recover a dependence of one random variable Y on another random variable X, i.e., how to determine the unknown function

$$Y = f(X). \tag{5.1}$$

More precisely, let the data for the random variables X and Y be given in the form of observations presented in the following table

TABLE 5.1: Data table.

X	x_1	x_2	x_3	$\cdots\ \cdots$	x_n
Y	y_1	y_2	y_3	$\cdots\ \cdots$	y_n

How to recover the function f from the data? The formulation of this question is still incomplete since one knows that any function can be defined by a table, in this case, by the pairs (x_i, y_i) where $y_i = f(x_i)$ for $i = 1, 2, \ldots, n$. Such restoring of f has no sense because the next observation $y_{n+1} = f(x_{n+1})$ is just a definition of f in the next point and cannot be predicted. A type of the function f must be additionally described. For instance, in Sec.3.2, it is assumed that f is a polynomial of degree $n-1$. Then $f(x_{n+1})$ can be predicted as the value of the interpolating polynomial at x_{n+1}. However, an interpolation does not pass as a restoration problem when it is known, for instance, that the approximate polynomial is a linear function. Moreover, the interpolation is not a proper method for random data because randomness means that some pairs (x_i, y_i) represent the dependence f with small probability. Hence, some pairs can essentially disturb the sought dependence. So, let us define the type of f assuming that it is a linear function and treat the pairs (x_i, y_i) from the table as approximate data. Using that point of view we introduce the random error of observation ξ and consider the dependence

$$Y = a_0 + a_1 X + \xi, \tag{5.2}$$

where the unknown parameters a_0 and a_1 have to be chosen.

Fig.5.1 demonstrates a typical situation when a cloud of points is approximated by a line.

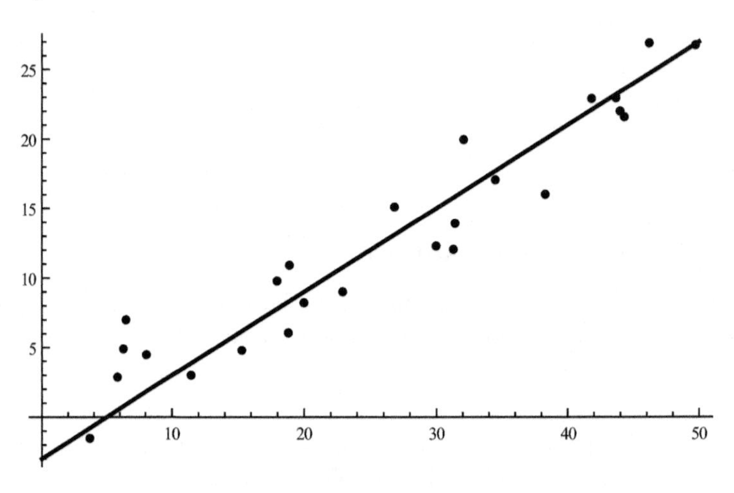

FIGURE 5.1: Line near a set of points

Substitution of the given pairs (x_i, y_i) from the table into (5.2) yields

$$y_i = a_0 + a_1 x_i + \xi_i, \quad i = 1, 2, \ldots, n. \tag{5.3}$$

We want the set of the calculated values $\xi_i = y_i - a_0 - a_1 x_i$, $(i = 1, 2, \ldots, n)$ to be "minimal" in some sense. One of the natural criteria of the "minimal set" is *the least squares* criterion when the sum of squares of ξ_i attains the minimal value:

$$\sum_{i=1}^{n} \xi_i^2 \to \min . \tag{5.4}$$

The linear equation

$$y = a_0 + a_1 x \tag{5.5}$$

constructed by use of the criterion (5.4) is called *the regression equation.*

Introduce the function $g(a_0, a_1)$ of two variables a_0 and a_1

$$g(a_0, a_1) = \sum_{i=1}^{n} (y_i - a_0 - a_1 x_i)^2. \tag{5.6}$$

Now, the minimization problem (5.4) is reduced to the standard problem of calculus, to find the minimal value of the function (5.6). Following the standard scheme we have to calculate the partial derivatives of (5.6) and equal them to zero

$$\frac{\partial g}{\partial a_0}(a_0, a_1) = 0, \quad \frac{\partial g}{\partial a_1}(a_0, a_1) = 0. \tag{5.7}$$

Simple computations yield the system of two linear algebraic equations for a_0 and a_1

$$\sum_{i=1}^{n}(y_i - a_0 - a_1 x_i) = 0, \tag{5.8}$$

$$\sum_{i=1}^{n} x_i(y_i - a_0 - a_1 x_i) = 0. \tag{5.9}$$

Substitute (5.3) into (5.8)-(5.9)

$$\sum_{i=1}^{n} \xi_i = 0, \quad \sum_{i=1}^{n} x_i \xi_i = 0. \tag{5.10}$$

These equations can be considered as consequences of the minimization problem (5.4).

In order to solve the system (5.8)-(5.9) introduce the mean value[1]

$$\overline{z} := \frac{1}{n}\sum_{i=1}^{n} z_i. \tag{5.11}$$

Then (5.8) becomes

$$\overline{y} = a_0 + a_1 \overline{x}. \tag{5.12}$$

It follows from (5.12) that the linear regression equation always passes through the point $(\overline{x}, \overline{y})$.

Equation (5.9) can be written in the form

$$\sum_{i=1}^{n} x_i y_i = a_0 \sum_{i=1}^{n} x_i + a_1 \sum_{i=1}^{n} x_i^2. \tag{5.13}$$

Substitute a_0 from (5.12) into (5.13)

$$\sum_{i=1}^{n} x_i y_i = (\overline{y} - a_1 \overline{x}) \sum_{i=1}^{n} x_i + a_1 \sum_{i=1}^{n} x_i^2. \tag{5.14}$$

Rewrite (5.14) in the form

$$\sum_{i=1}^{n} x_i y_i - n\overline{x}\,\overline{y} = a_1 \left(\sum_{i=1}^{n} x_i^2 - n\overline{x}^2 \right)$$

[1]In statistics, $\overline{z}$ is used for the mean value. In complex analysis, the bar means the complex conjugation.

and find

$$a_1 = \frac{\sum_{i=1}^{n} x_i y_i - n\overline{x}\,\overline{y}}{\sum_{i=1}^{n} x_i^2 - n\overline{x}^2} = \frac{\sum_{i=1}^{n}(x_i - \overline{x})(y_i - \overline{y})}{\sum_{i=1}^{n}(x_i - \overline{x})^2}. \tag{5.15}$$

Thus, the coefficient of the linear regression (5.5) is given by formulae (5.12) and (5.15).

We have found that the function $g(a_0, a_1)$ has a unique extremal value on $\mathbb{R}^2$. One can check that this extremum is the global minimum by use of the second derivatives. Another way is to look at the function (5.6) and just note that it is a non-negative parabolic function which has a unique minimum.

Though the best linear approximation has been constructed, it is not the complete solution to the problem because the dependence (5.1) can be essentially nonlinear and any linear approximation may give an unsatisfactory result. A degree of linear dependence can be measured by means of the following notation.

Definition 5.1. The value

$$r_{xy} = \frac{\sum_{i=1}^{n}(x_i - \overline{x})(y_i - \overline{y})}{\sqrt{\sum_{i=1}^{n}(y_i - \overline{y})^2 \sum_{i=1}^{n}(x_i - \overline{x})^2}}. \tag{5.16}$$

is called *the correlation coefficient* between the random variables X and Y with the realizations x_i and y_i, respectively $(i = 1, 2, \ldots, n)$.

The correlation coefficient satisfies the inequality $-1 \le r_{xy} \le 1$ and expresses a measure of the *linear dependence* between X and Y.

Definition 5.2. The value

$$\sigma_x = \sqrt{\frac{1}{n}\sum_{i=1}^{n}(x_i - \overline{x})^2}$$

is called *the standard deviation* of the random variable X (from its mean value $\overline{x}$).

Using these notations we rewrite (5.15) in the form

$$a_1 = r_{xy}\frac{\sigma_y}{\sigma_x}. \tag{5.17}$$

We have chosen the criterion of the least squares (5.4) to determine the optimal coefficient a_0 and a_1 in (5.5). Is it possible to consider other criteria? Sure. Instead of (5.4), one can consider, for instance, the minimization

$$\sum_{i=1}^{n} |y_i - a_0 - a_1 x_i| \to \min. \tag{5.18}$$

Solution to the minimization problem will give another linear function (5.5).

Why is the least squares method the most popular in applications? Because it has profound proofs from the physical point of view. The potential energy of the harmonic oscillator is given by (4.43), where x denotes the displacement. Therefore, the sum (5.6) can be considered as the energy of a set of material points and its minimum as the minimum of energy which is always attained at the equilibrium state. Hence, the minimization (5.4) determines the equilibrium state. It is worth noting that the minimization (5.18) is used in *the linear programming.*

Why is the minimum of energy used in econometrics? We do not know any reasonable answer to this question. Perhaps, by tradition. If a theory works in one domain, it can be applied in another one by analogy.

5.2 Fitting

As we know from the previous section the type of the dependence (5.1) has to be described. For instance, if we assume (or know from other sources) that the dependence must be parabolic, we determine the coefficients of the function

$$f(x) = a_0 + a_1 x + a_2 x^2 \tag{5.19}$$

by the same least squares method. It is convenient to introduce the basis of approximations. In the considered parabolic approximation the basis consists of the elements 1, x, x^2. The minimization problem can be treated as the determination of the best coefficients a_0, a_1, a_2 of the function $f(x)$ in the basis $\{1, x, x^2\}$. Since this basis is not complete in the space of continuous functions (not all continuous functions are represented in the form (5.19)), the least squares method gives an approximate solution for the coefficients a_0, a_1, a_2.

Mathematica contains the operator **Fit[data, {1, x, x²},x]**, which finds a linear combination for the **data** given in the form of **Table** by the basis **{1, x, x²}**. Consider an example with the data $\sin\frac{i}{10}$ $(i = 0, 1, \ldots, 30)$:

```
In[1]:= T = Table[Sin[i / 10], {i, 0, 30}] // N

Out[1]= {0., 0.0998334, 0.198669, 0.29552, 0.389418, 0.479426, 0.564642,
         0.644218, 0.717356, 0.783327, 0.841471, 0.891207, 0.932039,
         0.963558, 0.98545, 0.997495, 0.999574, 0.991665, 0.973848,
         0.9463, 0.909297, 0.863209, 0.808496, 0.745705, 0.675463,
         0.598472, 0.515501, 0.42738, 0.334988, 0.239249, 0.14112}

In[2]:= prosta = Fit[T, {1, x}, x]

Out[2]= 0.555401 + 0.00551705 x
```

This linear approximation can be improved by addition of the elements x^2 and x^3 to the basis.

```
In[3]:= wielomian2 = Fit[T, {1, x, x^2}, x]

Out[3]= -0.182035 + 0.139596 x - 0.00418998 x^2

In[4]:= wielomian3 = Fit[T, {1, x, x^2, x^3}, x]

Out[4]= -0.163527 + 0.133161 x - 0.00369512 x^2 - 0.0000103094 x^3
```

The results of approximations are presented in the following graph where the given data are shown with points and the fitting curves are presented with solid lines.

```
In[5]:= Show[ListPlot[T, PlotStyle → {Black, PointSize[Large]}],
         Plot[{prosta, wielomian2, wielomian3}, {x, 0, 31},
          PlotStyle → {{Thick, Black}}]]
```

Out[5]=

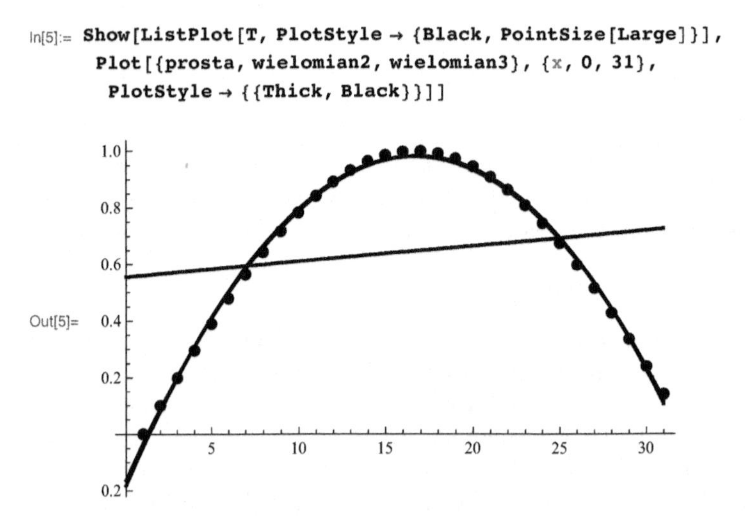

One can see that the linear approximation does not give good results and the approximations by polynomial of degree 2 and 3 almost coincide. Thus, the parabolic approximation by a polynomial of degree 2 give a good approximation. One can also analyze the outputs Out[3] and Out[4].

Consider a modified example when the given data are taken from an interval that is greater than the period of $\sin x$. This drastically changes the above approximations

```
In[6]:= T1 = Table[Sin[i / 10], {i, 0, 300, 10}] // N

Out[6]= {0., 0.841471, 0.909297, 0.14112, -0.756802, -0.958924, -0.279415,
         0.656987, 0.989358, 0.412118, -0.544021, -0.99999, -0.536573,
         0.420167, 0.990607, 0.650288, -0.287903, -0.961397, -0.750987,
         0.149877, 0.912945, 0.836656, -0.00885131, -0.84622, -0.905578,
         -0.132352, 0.762558, 0.956376, 0.270906, -0.663634, -0.988032}
```

The results of approximations by polynomials of degree 2, 3 and 15 are presented below.

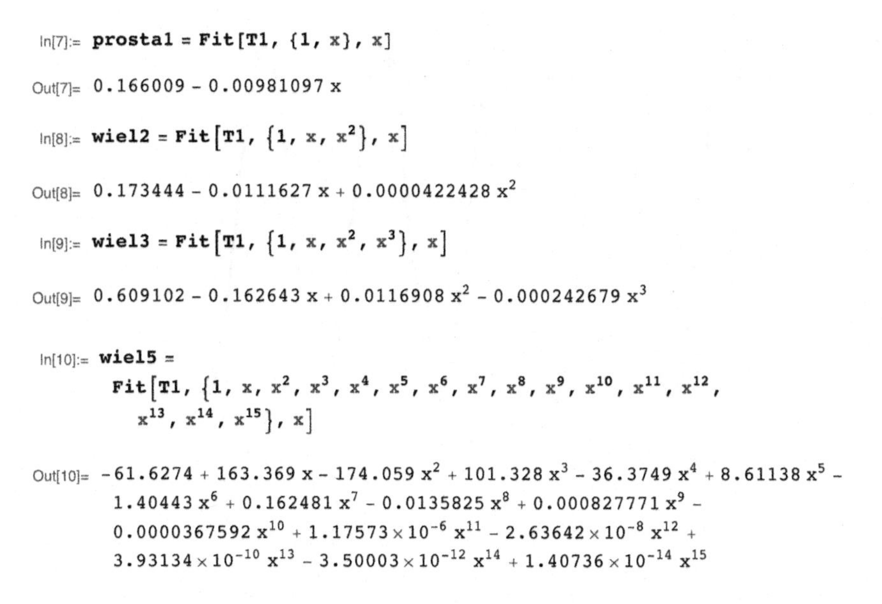

```
In[7]:= prostal = Fit[T1, {1, x}, x]

Out[7]= 0.166009 - 0.00981097 x

In[8]:= wiel2 = Fit[T1, {1, x, x^2}, x]

Out[8]= 0.173444 - 0.0111627 x + 0.0000422428 x^2

In[9]:= wiel3 = Fit[T1, {1, x, x^2, x^3}, x]

Out[9]= 0.609102 - 0.162643 x + 0.0116908 x^2 - 0.000242679 x^3

In[10]:= wiel5 =
          Fit[T1, {1, x, x^2, x^3, x^4, x^5, x^6, x^7, x^8, x^9, x^10, x^11, x^12,
            x^13, x^14, x^15}, x]

Out[10]= -61.6274 + 163.369 x - 174.059 x^2 + 101.328 x^3 - 36.3749 x^4 + 8.61138 x^5 -
          1.40443 x^6 + 0.162481 x^7 - 0.0135825 x^8 + 0.000827771 x^9 -
          0.0000367592 x^10 + 1.17573×10^-6 x^11 - 2.63642×10^-8 x^12 +
          3.93134×10^-10 x^13 - 3.50003×10^-12 x^14 + 1.40736×10^-14 x^15
```

The results are presented in the graph:

```
In[11]:= Show[ListPlot[T1, PlotStyle → {Black, PointSize[Large]}],
          Plot[{prostal, wiel2, wiel3, wiel5}, {x, 0, 31},
           PlotStyle → {{Thick, Black}}, PlotRange → {-1.5, 1.5}]]
```

Out[11]=

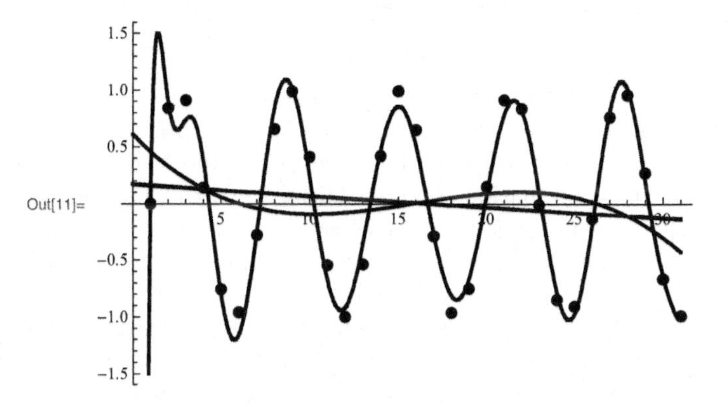

One can see that the polynomial approximation gives acceptable results for high degrees. It is not surprising that the trigonometric approximation of $\sin x$ has an excellent result:

```
In[12]:= approx = Fit[T1, {1, Sin[x], Cos[x]}, x]

Out[12]= -3.97888×10^-17 - 0.841471 Cos[x] + 0.540302 Sin[x]

In[13]:= Show[ListPlot[T1, PlotStyle → {Black, PointSize[Large]}],
          Plot[approx, {x, 0, 31}, PlotStyle → {{Thick, Black}},
           PlotRange → {-1.5, 1.5}]]
```

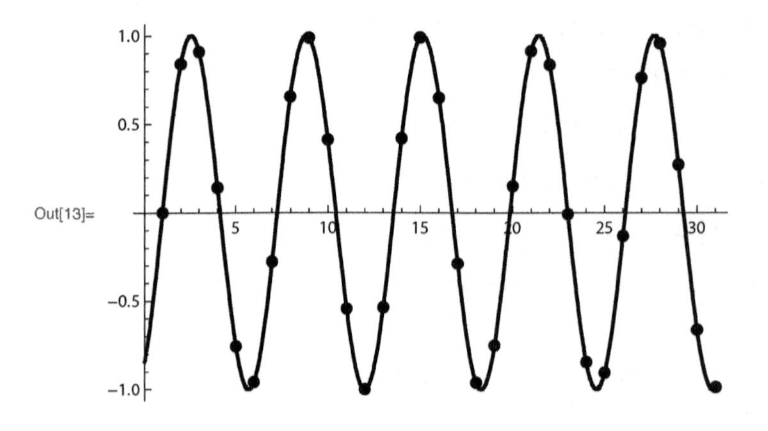

This observation leads us to

Principle of an apple. *An apple falls near its apple tree.*

Theorem 5.1 (Pinkus [51]). *Let $q \geq 1$ be an integer, $K \subset \mathbb{R}^q$ be a compact, $f : K \to \mathbb{R}^q$ be continuous. The function $\sigma : \mathbb{R}^q \to \mathbb{R}^q$ be continuous, but not polynomial. Then for every $\varepsilon > 0$ there exist $N \geq 1$, $a_k, b_k \in \mathbb{R}$, $\mathbf{w}_k \in \mathbb{R}^q$ such that*

$$\max_{\mathbf{x} \in K} \left| f(\mathbf{x}) - \sum_{k=1}^{N} a_k \sigma(\mathbf{w}_k \cdot \mathbf{x} + b_k) \right| < \varepsilon. \tag{5.20}$$

This theorem helps us to find an apple $\sum_{k=1}^{N} a_k \sigma(\mathbf{w}_k \cdot \mathbf{x} + b_k)$ near the apple tree $f(\mathbf{x})$. The function $\sigma(\mathbf{x})$ should be chosen by guess and the parameters a_k, b_k, $\mathbf{w}_k$ are found by fitting. For instance, if we know that $f(\mathbf{x})$ is related to a harmonic type oscillation (see Sec.4.6.1), we try to use periodic functions, namely, the trigonometric functions $\cos(ax + b)$ and $\sin(ax + b)$ as the components of $\sigma(\mathbf{x})$. The following elementary functions (composed with the linear function $y = ax + b$) are frequently taken as the test functions: logarithm, polynomial, rational functions, trigonometric, exponential, power functions x^α and their linear and non-linear combinations. In market and financial analysis, long and short time scales are applied. For instance, the test function $y = c + d\sqrt{t} + \varepsilon \cos(\nu t + \phi)$ for $d \gg \varepsilon$ models the main trend $c + d\sqrt{t}$ perturbed by the oscillating term $\varepsilon \cos(\nu t + \phi)$.

Mathematica possesses operators for nonlinear fits. Consider random data

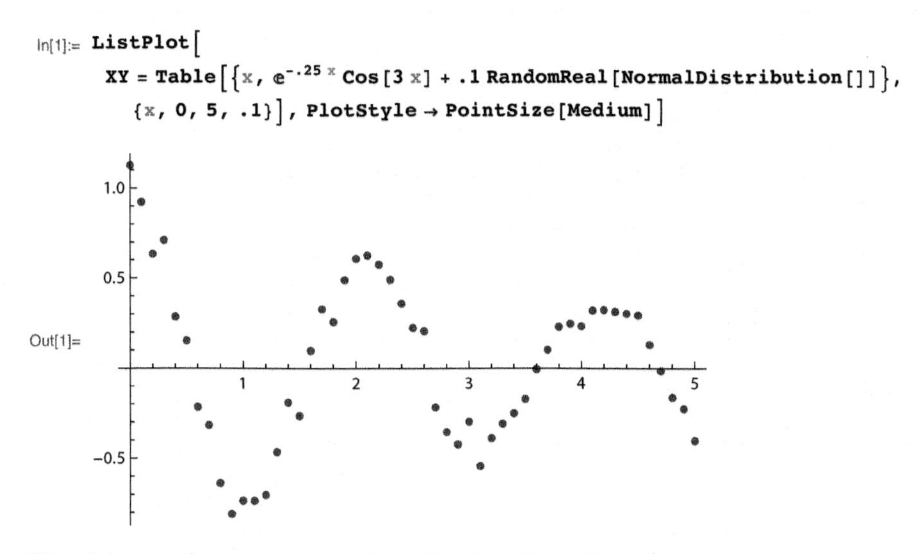

The data can be approximated by the function $e^{ax} \cos bx$:

```
In[2]:= F =
  (Model = NonlinearModelFit[XY, ⅇ^(a x) Cos[b x], {{a, 0}, {b, 1}},
     x, Method → Automatic]) ["BestFit"]

Out[2]= ⅇ^(-0.245213 x) Cos[3.01218 x]
```

The data in the considered example are artificially constructed. In practice, in accordance with the principle of an apples direct observations of various graphs and their comparisons with known samples can be useful in accurate approximations.

MATLAB Example Box 5.1

The following script uses `polyfit` operator performing polynomial curve fitting. The example shows two plots of polynomials of different orders. One can consider the model on the right-hand figure being too complex. Such situation is known as *overfitting*.

```
function script13()

    % sample perturbed data
    x_data = linspace(0, 2*pi, 12);
    eps = 0.2;
    y_data = 2*sin(x_data) + eps*cos(x_data.^4);

    % polynomial fitting
    p3 = polyfit(x_data, y_data, 3); % order 3
    p9 = polyfit(x_data, y_data, 9); % order 9
```

```
% plots of both polynomials
x = linspace(0, 2*pi);

subplot(1,2,1);
y = polyval(p3, x);
plot(x_data, y_data, 'o', x, y, 'k')
axis equal
axis([0 2*pi -2.2 2.2])

subplot(1,2,2);
y = polyval(p9, x);
plot(x_data, y_data, 'o', x, y, 'k')
axis equal
axis([0 2*pi -2.2 2.2])
```

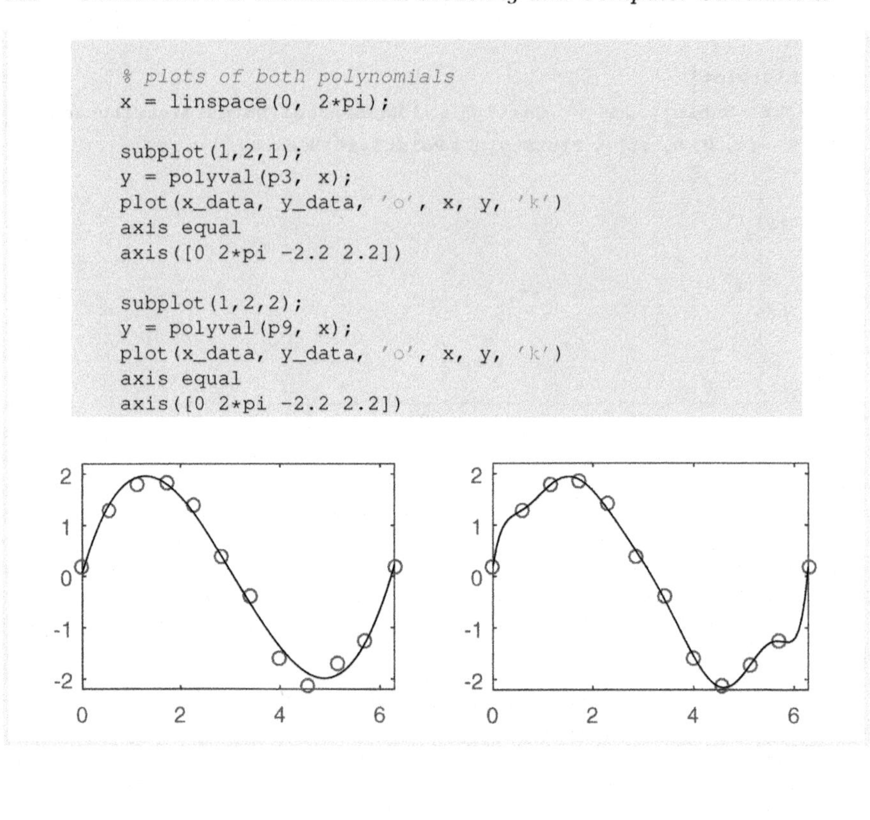

5.3 Method of Monte Carlo

The method of Monte Carlo is a simulation method based on the realizations of random variables. We now present an idea of this method on the example of computation of the number π, the ratio of the circle length to its diameter.

It is known that the area of the unit disk holds π. Hence, estimation of this area simultaneously yields the estimation of the number π. Let Ω be the square 2×2 and A be the disk enclosed to the square. We do the following experiment. Let n points be randomly thrown on the square. Here, randomness means that each point is represented by the coordinate (x_k, y_k) and the both coordinates x_k and y_k are randomly chosen in the interval $(-1, 1)$, i.e., they are uniformly distributed on this interval.

A part of these points, let it be m points, goes onto the disk. Let the number n be sufficiently large. Then, one can expect that the ratio $\frac{m}{n}$, called in statistics the frequency of the considered event, will be close to the ratio of the areas of the disk to the area of the square. In the considered case, it is equal to $\frac{\pi}{4}$. Therefore, the number $\frac{4m}{n}$ will approximate the sought value of π.

This is our strategy which can be realized by *Mathematica* as follows. First, take the number of points

```
In[1]:= n = 2000;
```

Second, we simulate n-multiple random point in the square Ω by two random coordinates with each of them being uniformly distributed on the interval of the length 2 (see picture on Fig.5.2).

```
In[2]:= pkt = Table[{RandomReal[{-1, 1}], RandomReal[{-1, 1}]}, {n}];

In[3]:= Graphics[{{EdgeForm[Dashed], White, Rectangle[{-1, -1}, {1, 1}]},
          {EdgeForm[Dashed], Lighter[Gray, 0.75], Disk[]},
          Darker[Gray, .1], Point /@ pkt, Black}]
```

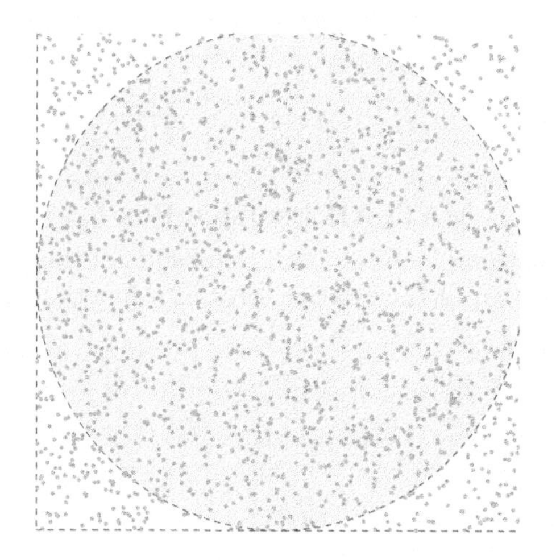

FIGURE 5.2: Square filled with the cloud of random points.

The operator **Select** deletes points outside the disk.

```
In[4]:= pkt = Select[pkt, #[[1]]^2 + #[[2]]^2 ≤ 1 &];

In[5]:= Graphics[{{EdgeForm[Dashed], White, Rectangle[{-1, -1}, {1, 1}]},
          {EdgeForm[Dashed], Lighter[Gray, 0.75], Disk[]},
          Darker[Gray, .1], Point /@ pkt, Black}]

In[6]:= dA = Length[pkt]

Out[6]= 1563
```

Therefore, the value of π and the absolute error have the form

```
In[7]:= P = 4 dA/n // N

Out[7]= 3.126

In[8]:= N[Abs[π - P], 10]

Out[8]= 0.0155927
```

The above algorithm was presented step by step and can be gathered into one operator:

```
In[9]:= pmc[n_] :=
          {P =
            Length[Select[Table[RandomReal /@ {{-1, 1}, {-1, 1}}, {n}],
              #[[1]]^2 + #[[2]]^2 ≤ 1 &]] 4/n, Abs[π - P]} // N
```

Perform computations for $n = 25000$ and $n = 100000$.

```
In[10]:= pmc /@ {25 000, 100 000}

Out[10]= {{3.1688, 0.0272073}, {3.14032, 0.00127265}}
```

MATLAB Example Box 5.2

The present example draws both the results of the Monte Carlo simulation and the plot of the relative error as a function of the number of samples.

```
function script14()

    % generate random data with n samples
    n = 3000;
    x = random('unif', 0, 1, 1, n);
    y = random('unif', 0, 1, 1, n);

    % compute logical vector
    accepted = x.^2 + y.^2 < 1;
    % compute vector of errors
    k = 1:n;
    errors = abs(pi - 4*cumsum(accepted) ./ k) / pi;

    % plot results
    subplot(1, 2, 1);
    plot(x(accepted), y(accepted), '.k', ...
        x(~accepted), y(~accepted), '.g')
    axis equal
    axis([0 1 0 1])

    subplot(1,2,2);
    plot(k, errors, '.k')
    axis square
    grid on
```

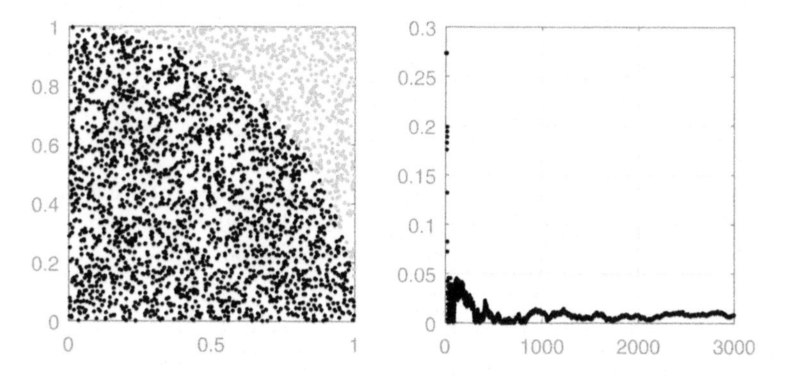

Here, the `accepted` array is a *logical vector*, i.e. the vector consisting of ones and zeros indicating whether the element on a given place meets the condition or not. The logical vector is used to reference of subarrays of points that were accepted (`x(accepted)`) as well as those rejected (applying the logical negation in `x(~accepted)`).

5.4 Random walk

Robert Brown (1828) studied motion of particles named after him Brownian motion. He tried to find a "primitive life force" and "active molecules". The independent papers by Albert Einstein (1905) and by Marian Smoluchowski (1906) can be considered as a rigorous demonstration of Brownian motion based on the diffusion equation (see Chapter 8, Remark 8.2 on page 172) and random walks (paths with random steps).

Simple 1D random walk is modeled by random steps to the left or to the right on the x-axis. In order to describe the model we recall elementary notations concerning discrete random values. Let X denote a random variable which takes a value x_i with a probability p_i ($i = 1, 2, \ldots, n$). The random value X can be introduced by Table 5.2 where x_i are real numbers, $0 < p_i \leq 1$ and $\sum_{i=1}^{n} p_i = 1$.

TABLE 5.2: Description of the random value X.

X	x_1	x_2	x_3	$\cdots\ \cdots$	x_n
P	p_1	p_2	p_3	$\cdots\ \cdots$	p_n

The mathematical expectation is defined as

$$E[X] = \sum_{i=1}^{n} p_i x_i. \tag{5.21}$$

The expectation (5.21) can be written in terms of the statistical mean value (5.11) as $E[X] = \overline{X}$. It gives the average value which can be expect in a realization of X. The variance is defined by formulae

$$Var[X] = E[(X - E[X])^2] = \sum_{i=1}^{n} p_i (x_i - E[X])^2 = \sum_{i=1}^{n} p_i x_i^2 - E[X]^2. \tag{5.22}$$

It shows the squared deviations from the expected value $E[X]$. If $E[X] = 0$, the variance $Var[X]$ can be written as $\overline{X^2}$.

We now proceed to discuss the 1D random walk. Time can be introduced as the number of steps made by a walker. Let the walker be at the point $x = x_0$ at time n. He/she can go to $x = x_0 - 1$ or to $x = x_0 + 1$ with the same probability $\frac{1}{2}$. Let every step at each time $n = 0, 1, 2, \ldots$ be independent of the previous steps. Therefore, we have the set of random mutually independent variables $X_1, X_2, \ldots$. Every random variable X_i takes two values -1 and $+1$ with the probability $\frac{1}{2}$. Let the walker be at the origin $x = 0$ at the initial time $n = 0$. We are interested in the position S_n of the walker at time n equal to the sum of all the previous steps

$$S_n = X_1 + X_2 + \ldots + X_n. \tag{5.23}$$

The mathematical expectation of the random variable S_n can be calculated as follows

$$E[S_n] = E[X_1 + X_2 + \ldots + X_n] = E[X_1] + E[X_2] + \ldots + E[X_n]. \tag{5.24}$$

Using formula (5.21) we obtain $E[X_i] = \frac{1}{2}(-1) + \frac{1}{2}(+1) = 0$ for any i. Therefore, the expected location of the walker is $x = 0$ for any time n. The same result follows from the symmetry of the random walks[2]. This first result is clear and true, but useless.

We go ahead and calculate the variance

$$Var[S_n] = E[(X_1 + X_2 + \ldots + X_n)^2] = \sum_{i=1}^{n} E[X_i^2] + \sum_{i=1}^{n} \sum_{k \neq i} E[X_i X_k]. \tag{5.25}$$

Always $X_i^2 = 1$, hence $E[X_i^2] = 1$. For fixed i and k, the random variable $X_i X_k$ takes the values -1 and $+1$ with the probability $\frac{1}{2}$. Therefore, all the terms in the double sum of (5.25) cancel each other out by pairs. Then, (5.25) becomes

$$Var[S_n] = n \iff \overline{S_n^2} = n. \tag{5.26}$$

[2]The symmetry principle by Curie (1894) states that symmetric causes imply symmetric effects.

This result means that the expected deviation of S_n^2 from the origin is equal to n. Therefore, the walker should be expected at the distance about $\sqrt{n}$ from the origin after n steps.

Formula (5.26) has been deduced for the integer scaling of space and time. Introducing another scale we can write (5.26) in the form

$$\overline{r^2(t)} = 2Dt, \tag{5.27}$$

where $r(t)$ denotes the distance between the walker (particle) and the origin at time t, the proportionality coefficient D is called the *diffusion constant.* The constant D expresses the mobility of the particle. In physics, it is proportional to the absolute temperature, in particular, D vanishes for the absolute zero (frozen particle). It is worth noting that equation (5.27) shows the average dependence for many particles, not a single one. The distance $r(t)$ for a fixed particle is an unpredictable value. But the average $r^2(t)$ of sufficiently many particles can be estimated by (5.27). Nevertheless, equation $r^2(t) \approx 2Dt$ holds for a single particle for sufficiently large t. Such a feature of the random (stochastic) processes is called *ergodicity.* An ergodic process has the same average properties for many particles in space and for a single particle in time.

Similar arguments can be used in the multidimensional spaces $\mathbb{R}^2$ and $\mathbb{R}^3$ to investigate the random walk on the integer lattices $\mathbb{Z}^2$ and $\mathbb{Z}^3$. The main result has the same form (5.26). The corresponding scaled formulae become

$$\overline{r^2(t)} = 4Dt \text{ in } \mathbb{R}^2; \quad \overline{r^2(t)} = 6Dt \text{ in } \mathbb{R}^3. \tag{5.28}$$

Here, $r(t) = |\mathbf{x}(t)|$ where $\mathbf{x}(t)$ denotes the spatial coordinate of the particle.

We now investigate the trajectory of the particle L when the time interval between steps tends to zero, i.e., the particle collides with other particles very frequently. For definiteness, consider 2D random walks. The trajectory L is a complicated zigzag continuous curve (see Fig.5.3) which consists in small segments Δl_i. All the segments has the same length $|\Delta l| = |\Delta l_i|$ but their inclinations are modeled by the random variable uniformly distributed in $(0, 2\pi)$. Introduce such a scale of distances $|\Delta \ell|$ that $|\Delta l| \ll |\Delta \ell| \ll r(t)$. The first equation (5.28) in this scale for one particle becomes

$$|\Delta \ell| = 2\sqrt{D\Delta t}. \tag{5.29}$$

Let the segment $\Delta \ell$ be chosen on the curve L with a fixed its beginning point. Equation (5.29) can be written in the form

$$\frac{|\Delta \ell|}{\Delta t} = \frac{2\sqrt{D}}{\sqrt{\Delta t}}. \tag{5.30}$$

The limit

$$\lim_{\Delta t \to 0} \frac{|\Delta \ell|}{\Delta t} \tag{5.31}$$

should yield the instant velocity of the particle along the limit curve L. However, it does not exist. In particularly, this means that the limit curve L is continuous everywhere but smooth nowhere (there is a cusp at any point of the curve), i.e., the parametric equations of the curve are described by functions differentiable nowhere. At the same time,

$$\lim_{\Delta t \to 0} \frac{|\Delta \ell|}{(\Delta t)^{\frac{1}{2}}} = 2\sqrt{D}. \tag{5.32}$$

The latter limit can be considered as the fractional derivative of order $\frac{1}{2}$. This leads us to the theory of fractals [24]. We will return to Brownian motion at the end of Sec.5.4 (see Remark 8.2) studying the diffusion (heat) equation. A formal mathematical introduction to the theory of discrete random variables can be found in [16, 37].

An example of the codes to simulate the 2D Brownian motion of a particle is given below.

```
In[1]:= M = 1000;

In[2]:= θ = 2 π RandomReal[1, M];

In[3]:= a[0] = {0, 0}; a[n_] := Sum[{Cos[θ[[k]]], Sin[θ[[k]]]}, {k, 1, n}]

In[4]:= Graphics[Line[Table[a[n], {n, 0, M}]]]
```

The graphs in Fig.5.3 are based on the above code.

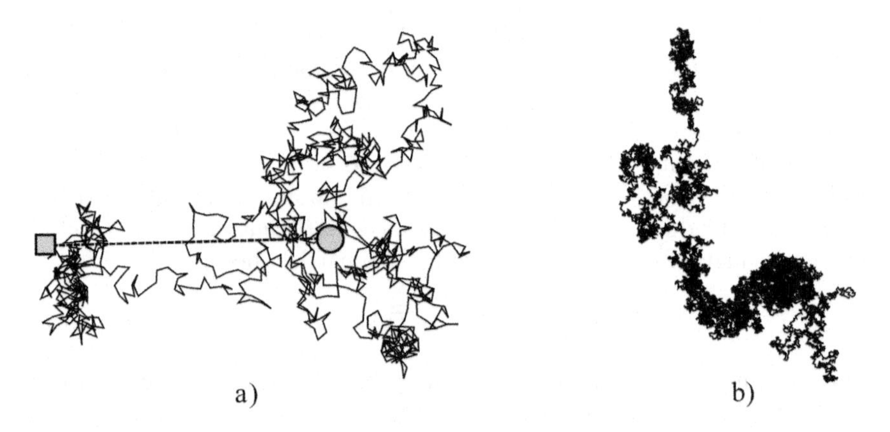

FIGURE 5.3: 2D Brownian motion. a) The original location (disk) is connected to the ultimate location (square) after 1000 random steps. b) 10000 random steps. The linear size is 10 times less than in the figure (a).

MATLAB Example Box 5.3

In a similar way one can simulate random walk in MATLAB.

```
function script15()

    % number of steps
    n = 1000;

    % generate directions and compute coordinates
    directions = random('unif', 0, 2*pi, 1, n);
    x = cumsum(cos(directions));
    y = cumsum(sin(directions));

    % plot the path
    plot(x, y, '.-k')
    axis equal
    grid on
```

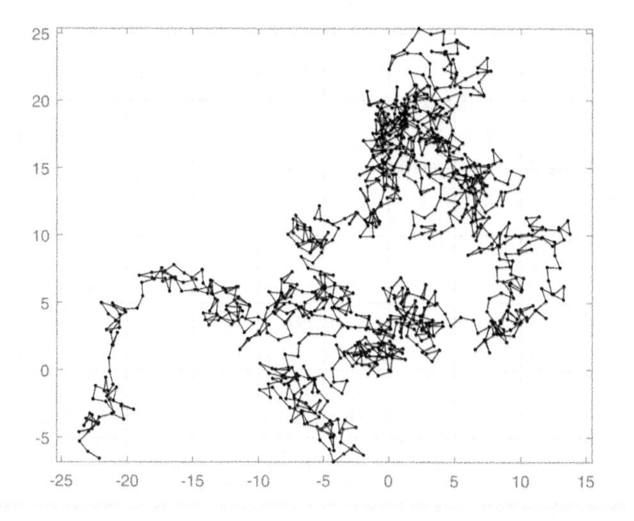

Concluding remarks and further reading. In the present chapter, we are concerned with statistical methods highly developed for various applications. For computational methods in statistics we recommend the textbook [58] based on the open-source software package *R*. Introduction to stochastic methods is presented in [37].

Exercises

1. Using the operator **`FinancialData`** in *Mathematica* retrieve the data on U.S. stock for GE (General Electric) and BP (British Petroleum),

for instance from May 22, 2015. Fit the data by a linear function, by polynomials and by trigonometric functions separately for GE and for BP.

Calculate the correlation between GE and BP by formula (5.16).

Using the constructed fitting models try to predict the prices of GE and for BP.

Analyze the data on U.S. stock for various firms dividing them into groups (for instance energetic, computer etc)[3].

2. Compute the area of the domain bounded by the ellipse $\frac{x^2}{9} + \frac{y^2}{4} = 1$ applying the Monte Carlo simulations.

   Compute the area $S(a,b)$ of the domain bounded by the ellipse $\frac{x^2}{a^2} + \frac{y^2}{b^2} = 1$ applying the Monte Carlo simulations.

   Hint 1: Investigate numerically the dependence of $S(a,b)$ on two parameters, i.e., on a with a fixed b and on b with a fixed a.

   Hint 2: An alternative way is based on the formula $S(ka, kb) = k^2 S(a,b)$ where $k > 0$ is a linear extension coefficient. Take $k = (ab)^{-\frac{1}{2}}$. Then, $S(a,b) = ab\ S(d, d^{-1})$ with $d = a^{\frac{1}{2}} b^{-\frac{1}{2}}$. Investigate numerically the dependence of $S(d, d^{-1})$ on d.

   Answer: $S(a,b) = \pi ab$.

3. Compute the volume of the domain bounded by the ellipsoid $\frac{x^2}{9} + \frac{y^2}{4} + \frac{z^2}{4} = 1$ applying the Monte Carlo simulations.

4. Compute the volume of the domain bounded by the ellipsoid $\frac{x^2}{a^2} + \frac{y^2}{b^2} + \frac{z^2}{c^2} = 1$ applying the Monte Carlo simulations.
   Answer: the volume holds $\frac{4}{3}\pi abc$.

5. Compute the surface area $S(a,b,c)$ of an ellipsoid by fitting the parameter p in the approximate formula

$$S(a,b,c) \approx 4\pi \left(\frac{a^p b^p + a^p c^p + c^p b^p}{3} \right)^{\frac{1}{p}}.$$

   Answer: $p \approx 1.6$.

[3]Caution: one should be aware of the hazards associated with the use of defective economic mathematical models. Justin Fox said about it: "Presenting a model is like doing a card trick. Everybody knows that there will be some sleight of hand. There is no intent to deceive because no one takes it seriously", *bloomberg.com/view/articles/2015-05-20/nyu-professor-starts-war-on-mathiness-in-economics*.

Economists frequently reduce theorems like *If A is fulfilled, then B holds* to the assertion *B holds*. For instance, the assertions *Investments increases during economic growth* and *Investments increases* are different.

6. Consider data presented in the form of a set of dimensional numbers. Take only the first digits of these numbers. Generate a statistical distribution of the first digits $\{1, 2, \ldots, 9\}$ following Simon Newcomb (1881) for

   a) populations of countries,

   b) areas of countries,

   c) heights of mountains higher than 2 km.

   Investigate the same statistical distributions but in other bases of number systems.

   Investigate the same statistical distributions of first digits in stock exchanges taking prices of one stock in time.

7. Prove formulae (5.28) and check them experimentally.

8. Display 3D Brownian motion similar to 2D Fig.5.3. One can apply **`Accumulate`** operator in *Matematica* code similar to `cumsum` from the MATLAB Example Box 5.3.

9. Study the asymmetric 1D random walk when the probabilities of the steps to left and to right are different.

Chapter 6

One-dimensional stationary problems

6.1	1D geometry	113
6.2	Second order equations	116
6.3	1D Green's function	120
6.4	Green's function as a source	123
6.5	The δ–function	126
	Exercises	130

6.1 1D geometry

In order to develop a mathematical model, first, we answered the question "where" and "when" in Sec.1.1.1. We described space and time for the considered process. It was easy to introduce time increasing form the initial time t_0 to the end time t_1. Frequently, it is convenient to take $t_0 = 0$, sometimes $t_0 = -\infty$ and $t_1 = +\infty$. So, the time $t \in \mathbb{R}$ is always one dimensional. The space variable $\mathbf{x} = (x, y, z)$ is a vector of the three-dimensional (3D) space $\mathbb{R}^3$. However, as we saw before, sometimes the considered phenomenon is essentially 1D, i.e., depends only on one space variable, say x. The reduction of the 3D to the 1D geometry is displayed in Figs 6.1-6.2.

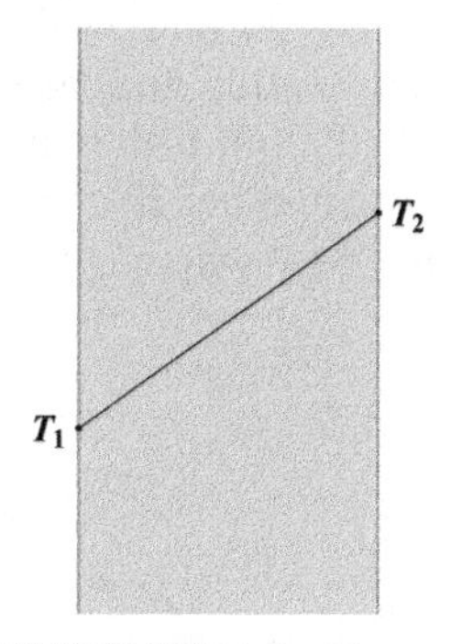

FIGURE 6.1: 1D temperature distribution in the 3D wall.

1D problems arise also for cylinder symmetry (Fig.6.3) or for spherical symmetry (Fig.6.4). In some examples, such a symmetry is not observed, but the process can be studied locally in the neighbourhood of a point P where the local field can be described by 1D geometry (Fig. 6.5).

We should use special units for 1D problem . For instance, the density of the 1D wire ρ_1 is measured in $\frac{kg}{m}$, though the standard density ρ_3 is measured in $\frac{kg}{m^3}$. The simplification: $\mathbb{R}^3 \to \mathbb{R}$, called in mathematics by projection, transforms the units: $m^3 \to m$ that gives the unit dependence $[\rho_3] = [\rho_1]m^{-2}$.

Another type of simplification is based on the elimination of time, if the

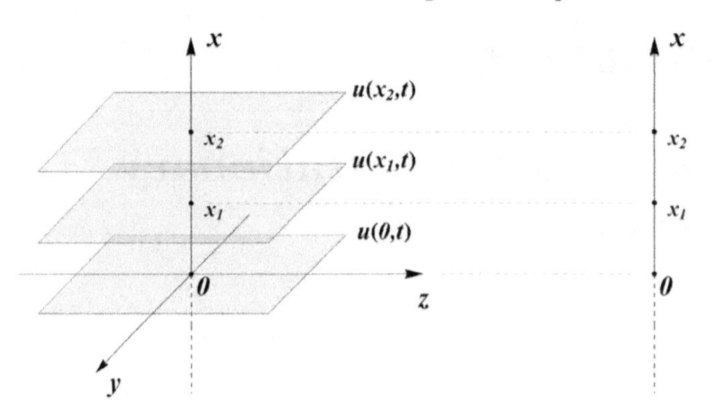

FIGURE 6.2: Let a function $u(x,t)$ is defined in the half-plane $\left\{(x,y,z) \in \mathbb{R}^3 : x > 0\right\}$ and does not depend on y and z. Then, the function $u(x,t)$ can be investigated as a function in the 1D space of variable x on the half-axis $x > 0$.

considered process does not change in time. Consider such a 1D stationary problem when the process does not depend on time. Actually, everything is going and observed in time, but frequently it is worth forgetting about it. The temperature distribution in the wall (see Fig.6.1) depends on the exterior temperature in the street T_1 and the interior temperature in the apartment T_2, but T_1 and T_2 do not change in the restricted interval of time. The heat flux steadily goes at all times because of the constant temperature difference $T_2 - T_1$, but it does not depend on t. In such cases, we may introduce the units measured per the time unit, for instance, stationary heat flux can be measured in $\frac{cal}{sec\ m^2}$ where cal denotes calorie ($1cal = 4.1855J$, J denotes Joule).

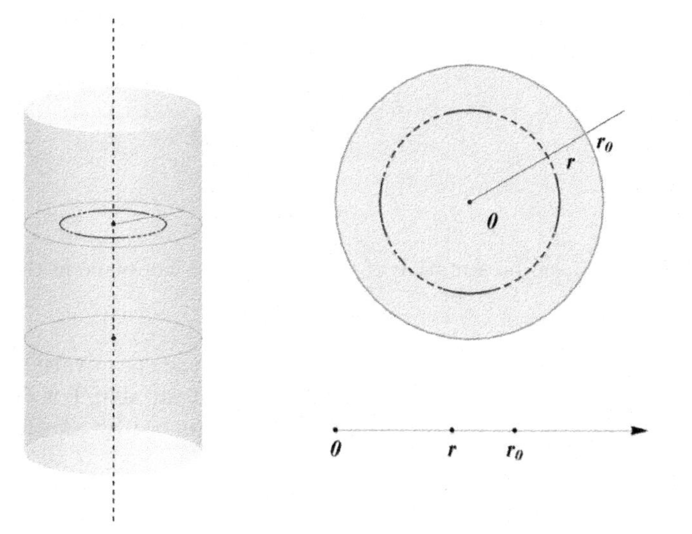

FIGURE 6.3: Let the temperature distribution be same in every section perpendicular to the cylinder axis. Then instead of the cylinder in $\mathbb{R}^3$ we can consider a disk in $\mathbb{R}^2$. Moreover, if the temperature distribution in every point of the disk depends only on the distance form the point O, then instead of the disk in $\mathbb{R}^2$ we can consider an interval $0 \le r \le r_0$, where r_0 denotes the radius of the disk.

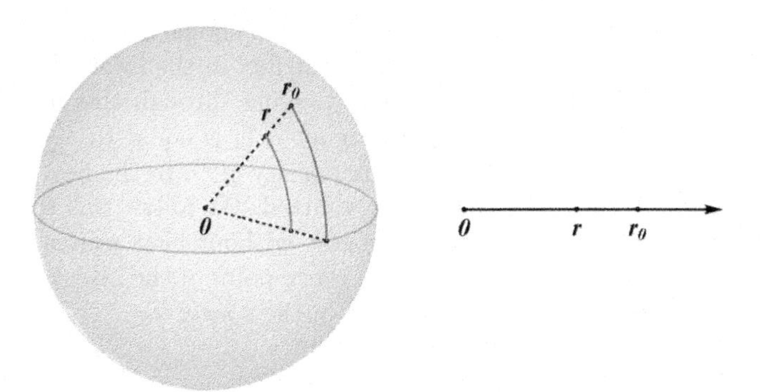

FIGURE 6.4: Let the temperature distribution in the ball of the radius r_0 be dependent only on the distance r to the center of the ball. Then, the temperature distribution depends only on $r \in [0, r_0]$.

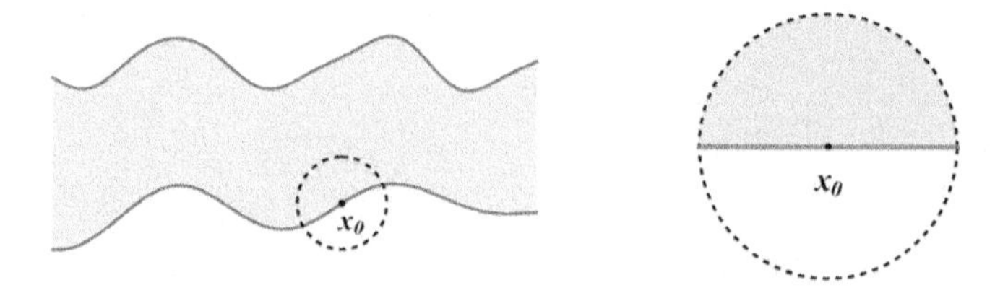

FIGURE 6.5: Local presentation of the domain as a half-plane in the vicinity of the point x_0.

This chapter is devoted to 1D problems when the space variable x lies in the axis $\mathbb{R}$. Time t can be considered as the second variable. For definiteness, we consider the heat conduction which can be replaced by another physical process, cf. Sec.8.7.

6.2 Second order equations

Consider 1D stationary heat conduction in the interval $[a, b]$. The fundamental physical values of the process are *the temperature distribution* given in the form of a function $u(x)$ defined in $[a, b]$ and *the heat flux density* in the form of a function $q(x)$ also defined in $[a, b]$. The heat flux density $q(x)$ is related to the heat flux. Let $Q_{\Delta x}$ be the heat flux in the interval $(x, x + \Delta x)$. More precisely, $Q_{\Delta x}$ is the value of heat passing through this interval per a time unit. Therefore, the mean heat flux at $[a, b]$ is equal to $\frac{Q_{\Delta x}}{\Delta x}$. Going to the limit $\Delta x \to 0$ we obtain the heat flux density $q(x)$. In the heat conduction theory the term "density" is frequently omitted. We follow this terminology.

At the beginning, we consider a process without heat sources and sinks, i.e., heat does not arise and disappear in every point of the interval $[a, b]$. This yields the conservation of the heat flux q in the space:

$$\frac{dq}{dx} = 0 \quad \text{for} \quad x \in (a, b). \tag{6.1}$$

The Fourier experimental law (1822) says that the total heat (in the time unit) passing through an interval of the length Δx is proportional to the temperature difference $\Delta u = u(x + \Delta x) - u(x)$

$$Q_{\Delta x} = -\lambda \Delta u. \tag{6.2}$$

The constant λ is called the heat conduction coefficient and depends on the type of material. The coefficient λ is always positive; the sign minus in (6.2) shows that heat always goes from hot to cold. The law (6.2) has the discrete form. Dividing both sides of (6.2) by Δx and going to the limit $\Delta x \to 0$ we arrive at the Fourier law in the differential form (cf. principle of transition: *continuous* $\leftrightarrow$ *discrete* on page 19)

$$q = -\lambda \frac{du}{dx}. \tag{6.3}$$

Equations (6.1), (6.3) constitute the fundamental laws of the 1D stationary heat conduction. Substituting (6.3) into (6.1) yields

$$\frac{d^2u}{dx^2} = 0 \quad \text{for} \quad x \in (a, b), \tag{6.4}$$

since λ is a constant.

Remark 6.1. *The interval (a, b) is open in equation (6.4) and $u \in C^1[a, b] \cap C^2(a, b)$, i.e., the function u is continuously differentiable in the closed interval $[a, b]$ and twice continuously differentiable in the open interval (a, b). These properties of u are caused by the condition for the heat flux $q \in C[a, b] \cap C^1(a, b)$. The law (6.1) yields the condition $q \in C^1(a, b)$. The law (6.1) could not be fulfilled at the ends of the interval. Left- and right-sided continuity of the heat flux at the ends of the interval is based on the physical condition of the heat balance at the end points.*

General solution of the ODE (6.4) has the form

$$u(x) = C_1 x + C_2, \tag{6.5}$$

where C_1 and C_2 are arbitrary constants. This follows from the fact that only linear functions (6.5) have a vanishing second derivative. Equation (6.5) implies that the stationary temperature distribution in the wall is always linear as shown in Figs 6.6 and 6.1.

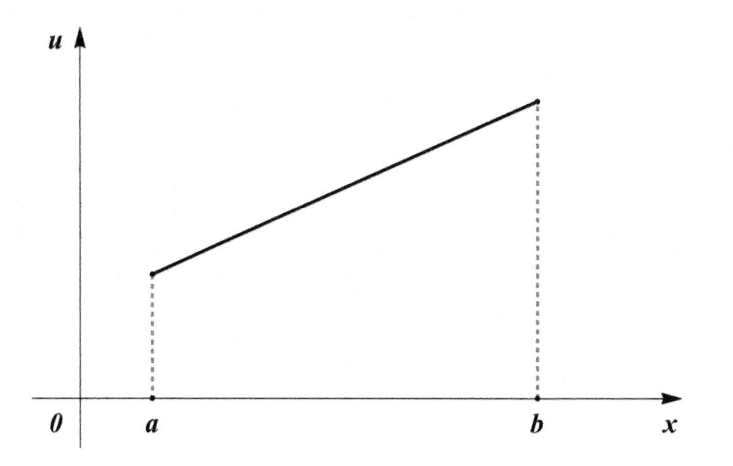

FIGURE 6.6: Linear temperature distribution u in the interval $[a, b]$.

We now proceed to discuss a medium which contains heat sources and sinks modeled by the density function $f(x)$[1]. Equation (6.1) can be modified as follows

$$\frac{dq}{dx} = f(x) \quad \text{for} \quad x \in [a,b]. \tag{6.6}$$

The given source function f is bounded in the interval $[a,b]$. Moreover, it is assumed that the function f is continuous in $[a,b]$ except a finite number of points where f has one-sided continuous limits. The total heat arisen in the interval $(c,d) \subset (a,b)$ per time unit can be estimated as the area between the graph of the source function $f(x)$ in (c,d) and the x-axis.

Substitution of (6.3) into (6.6) yields the ODE on the temperature distribution

$$\frac{d^2u}{dx^2} = -h(x) \quad \text{for} \quad x \in (a,b), \tag{6.7}$$

where $h(x) = \frac{f(x)}{\lambda}$. The boundary conditions at the points $x = a$ and $x = b$ has to be given in order to uniquely determine the temperature distribution $u(x)$.

Dirichlet boundary condition. Let the temperature be given at the boundary points, i.e., given the values α and β for which

$$u(a) = \alpha, \quad u(b) = \beta. \tag{6.8}$$

The triple $\{h, \alpha, \beta\}$ forms the Dirichlet data (6.7)–(6.8). One can see that these equations describe the linear heat conduction (c.f. the linear operator on page 19). Therefore, it is sufficient to solve two linear problems for the data $\{h, 0, 0\}$, $\{0, \alpha, \beta\}$ and add the results.

Theorem 6.1 (existence and uniqueness of solution). *Let the function h be continuous in $[a,b]$. Then, the Dirichlet problem* (6.7)–(6.8) *always has one and only one solution $u \in C^1[a,b] \cap C^2(a,b)$.*

The proof can be found for example in the book [55].

Equation (6.7) can arise in applications in more general classes of functions. Let the function h from the right-hand side of (6.7) be continuous in $[a,b]$ except a point x_0 for which $h(x_0+0) \neq h(x_0-0)$. Then, $u''(x_0+0) \neq u''(x_0-0)$, hence the function u belong to $C^1[a,b] \cap (C^2(a,x_0) \cap C^2(x_0,b))$.

Definition 6.1. The function

$$H(x) = \begin{cases} 0, & x < 0, \\ 1, & x > 0, \end{cases} \tag{6.9}$$

is called Heaviside's function.

[1] Such a case takes place during a chemical reaction going simultaneously with the heat conduction. This is a simplified model of the heat conduction coupled with the mass transfer [22].

Example 6.1. Let a point x_0 belongs to $(0,1)$. Consider the Dirichlet problem for equation (6.7) in the interval $(0,1)$ with $\alpha = \beta = 0$, $h(x) = H(x - x_0)$ where H is Heaviside's function. Then, equation (6.7) becomes

$$\frac{d^2u}{dx^2} = 0 \quad \text{for} \quad x \in (0, x_0) \tag{6.10}$$

and

$$\frac{d^2u}{dx^2} = -1 \quad \text{for} \quad x \in (x_0, 1). \tag{6.11}$$

Twice integration of equation (6.10) yields

$$u = C_1 x + C_3 \quad \text{for} \quad x \in (0, x_0).$$

It follows from the condition $u(0) = 0$ that $C_3 = 0$. Hence,

$$u = C_1 x \quad \text{for} \quad x \in [0, x_0). \tag{6.12}$$

Along similar lines

$$u = -\frac{(x-1)^2}{2} + C_2(1-x) \quad \text{for} \quad x \in (x_0, 1]. \tag{6.13}$$

The function $u'(x)$ must be continuous in $[0,1]$, in particular, at the point x_0:

$$u'(x_0+) = u'(x_0-). \tag{6.14}$$

The relation (6.14) expresses the continuity of the heat flux at the point x_0. This is also confirmed by the energy conservation at x_0. The continuity of the temperature at x_0 implies that

$$u(x_0+) = u(x_0-). \tag{6.15}$$

Two equations (6.14)–(6.15) constitute the linear system of equations on C_1 and C_2

$$\begin{cases} C_1 = 1 - x_0 - C_2, , \\ C_1 x_0 = -\frac{1}{2}(x_0 - 1)^2 + C_2(1 - x_0). \end{cases} \tag{6.16}$$

Solve this system

```
In[2]:= Solve[{C1 == 1 - x0 - C2, C1 x0 == -1/2 (x0 - 1)^2 + C2 (1 - x0)}, {C1, C2}]

Out[2]= {{C1 -> 1/2 (1 - 2 x0 + x0^2), C2 -> 1/2 (1 - x0^2)}}
```

This yields the temperature distribution

$$u(x) = \begin{cases} \frac{(1-x_0)^2}{2} x, & x \in [0, x_0], \\ -\frac{(1-x)^2}{2} + \frac{1-x_0^2}{2}(1-x), & x \in (x_0, 1]. \end{cases} \tag{6.17}$$

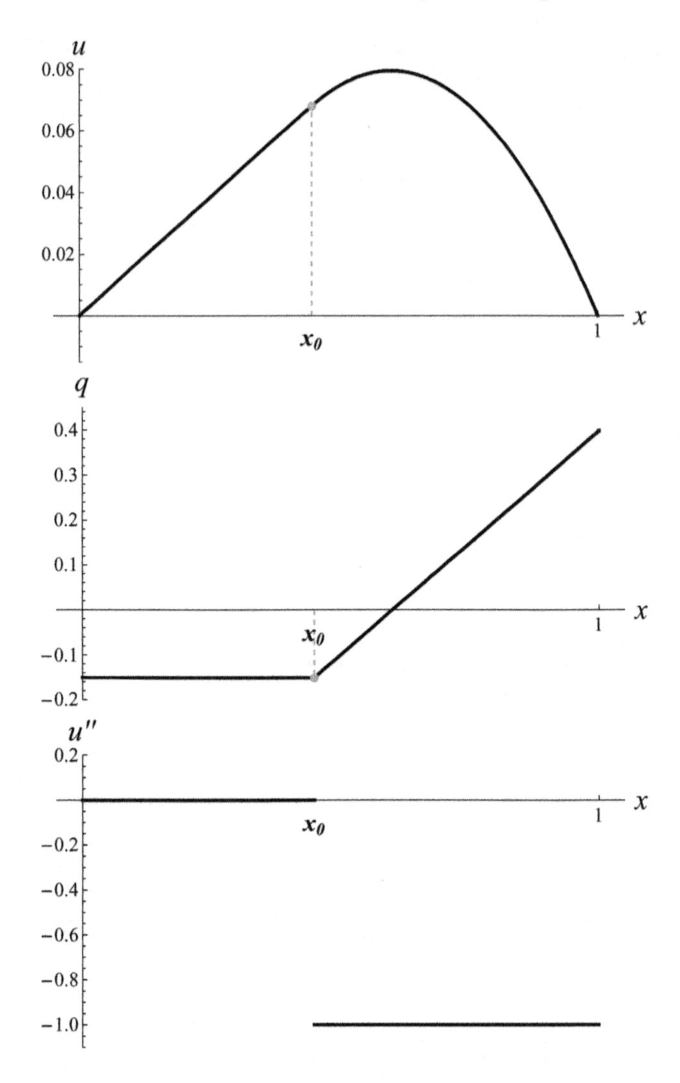

FIGURE 6.7: The temperature distribution u, the heat flux $q = -\lambda u'$ and the functions $u''(x) = -H(x - x_0)$. The data are for $\lambda = 1$.

6.3 1D Green's function

Following [55] we consider the problem (6.7)–(6.8) for the data $\{h, 0, 0\}$

$$\frac{d^2u}{dx^2} = -h(x) \quad \text{for} \quad x \in (a, b), \tag{6.18}$$

$$u(a) = 0, \quad u(b) = 0. \tag{6.19}$$

Multiply equation (6.18) by $(x - a)$ and calculate the integral

$$-\int_a^b (x-a)u''(x)dx = \int_a^b (x-a)h(x)dx.$$

Hence,

$$\int_a^b (x-a)u''(x)dx + \int_a^b (x-a)h(x)dx = 0. \tag{6.20}$$

The first integral (6.20) is calculated by parts. Using the boundary conditions (6.19) we obtain

$$\int_a^b (x-a)u''(x)dx = (x-a)u'(x)|_a^b - \int_a^b u'(x)dx = (b-a)u'(b).$$

Then, (6.20) becomes

$$(b-a)u'(b) + \int_a^b (x-a)h(x)dx = 0. \tag{6.21}$$

Come back to equation (6.18). Multiply it $(b - x)$ and repeat the above manipulations

$$-(b-a)u'(a) + \int_a^b (b-x)h(x)dx = 0. \tag{6.22}$$

Double integration of (6.18) from a to x yields

$$\begin{gathered} \frac{d^2u}{dx^2} = -h(x) \Leftrightarrow \int_a^x u''(\xi)d\xi = -\int_a^x h(\xi)d\xi \Leftrightarrow \\ u'(x) = u'(a) - \int_a^x h(\xi)d\xi \Leftrightarrow \\ u(x) = \int_a^x u'(\eta)d\eta = u'(a)(x-a) - \int_a^x d\nu \int_a^\eta h(\eta)d\eta. \end{gathered} \tag{6.23}$$

In order to transform the iterated integral from (6.23) to an ordinary integral we use Cauchy's formula

$$\int_a^x d\xi \int_a^\xi h(\eta)d\eta = \int_a^x (x-\xi)h(\xi)d\xi. \tag{6.24}$$

First, this formula is proved. Formula (6.24) holds for $x = a$. Differentiate equation (6.24) in x. The left-hand side becomes

$$L' = \int_a^x h(\xi)d\xi.$$

The right-hand side is reduced to the same expression because

$$P' = \left((x-a)\int_a^x h(\xi)d\xi - \int_a^x (\xi - a)h(\xi)d\xi\right)' = \int_a^x h(\xi)d\xi.$$

Thus, formula (6.24) is proved.

Application of (6.24) to (6.23) yields

$$u(x) = u'(a)(x-a) - \int_a^x (x-\xi)h(\xi)\, d\xi. \tag{6.25}$$

Substitute $u'(a)$ from (6.22) into (6.25) and make simple transformations of the integrals

$$u(x) = \frac{x-a}{b-a}\int_a^b (b-\xi)h(\xi)\, d\xi - \int_a^x (x-\xi)h(\xi)\, d\xi =$$

$$= \int_a^x \left[\frac{x-a}{b-a}(b-\xi) - (x-\xi)\right] h(\xi)\, d\xi + \int_x^b \frac{x-a}{b-a}(b-\xi)h(\xi)\, d\xi =$$

$$= \int_a^x \frac{(\xi-a)(b-x)}{b-a} h(\xi)\, d\xi + \int_x^b \frac{(x-a)(b-\xi)}{b-a} h(\xi)\, d\xi.$$

This gives

$$u(x) = \int_a^b g(x,\xi)h(\xi)d\xi, \tag{6.26}$$

where the function

$$g(x,\xi) = \begin{cases} \frac{1}{b-a}(\xi-a)(b-x), & a \le \xi \le x, \\ \frac{1}{b-a}(x-a)(b-\xi), & x \le \xi \le b, \end{cases} \tag{6.27}$$

is called *Green's function* for the problem (6.18)–(6.19). One can see that the structure of this function corresponds to the black box operator introduced on page 20. The operator in the right-hand side of (6.26) transforms the function h (input) to the function u (output). It is useful to check that the operator (6.26)-(6.27) is linear (see page 19).

Change the variables x and ξ in (6.27)

$$g(\xi,x) = \begin{cases} \frac{1}{b-a}(x-a)(b-\xi), & a \le x \le \xi, \\ \frac{1}{b-a}(\xi-a)(b-x), & \xi \le x \le b. \end{cases} \tag{6.28}$$

Comparing the first line of (6.27) and second line of (6.28) we conclude that

$$g(x,\xi) = g(\xi,x) \tag{6.29}$$

for $a \le \xi \le x \le b$. Similar comparison of the second line of (6.27) and the first line of (6.28) yields (6.29) for $a \leq x \leq \xi \leq b$. Therefore, formula (6.29) holds for all x and ξ from $[a, b]$ that means the symmetry of the function $g(x, \xi)$ on the variables.

The graph of $g(x, \xi)$ is displyed in Fig.6.8.

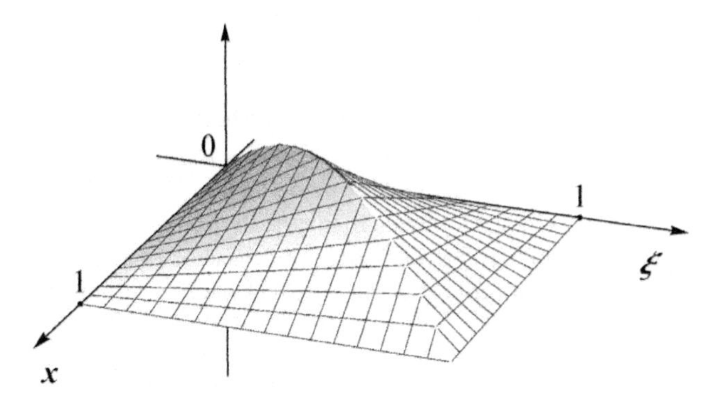

FIGURE 6.8: Graph of $g(x, \xi)$

The general problem

$$\frac{d^2u}{dx^2} = -h(x) \quad \text{dla} \quad x \in (a, b), \tag{6.30}$$

$$u(a) = \alpha, \quad u(b) = \beta, \tag{6.31}$$

has the unique solution

$$u(x) = \int_a^b g(x, \xi)h(\xi)d\xi + (b - x)\alpha + (x - a)\beta. \tag{6.32}$$

The integral in (6.32) is associated to the data $\{h, 0, 0\}$ and the linear function $(b - x)\alpha + (x - a)\beta$ solves the problem for the data $\{0, \alpha, \beta\}$. It follows from linearity that the sum of solutions for the data $\{h, 0, 0\}$ and $\{0, \alpha, \beta\}$ gives the solution (6.32) of the problem (6.30)-(6.31) for the data $\{h, \alpha, \beta\}$.

6.4 Green's function as a source

During development of the mathematical model we have not forgotten about the original problem (physical, economic etc) because eventually mathematical difficulties can be resolved by means of the pure physical level without direct mathematical methods. Moreover, a physical treatment of the problem helps to understand pure mathematical theories.

Principle of "trying" by Franklin D. Roosevelt. *If a mathematical problem is hard, it can be represented in a non-mathematical terms. "It is common sense to take a method and try it. If it fails, admit it frankly and try another. But above all, try something."*[2]*. Another look at the problem can lead to its solution.*

One can also stay in the framework of mathematics but try to represent the problem from a geometrical point of view. Geometry and visualization frequently help to solve non-geometric problems.

In this section, we apply physical intuition to model point singularities when the density attains (infinitely) large values. For definiteness, we will use the heat conduction terminology. Consider 1D heat conduction described by equations (6.18)–(6.19)

$$\frac{d^2u}{dx^2} = -h(x) \quad \text{for} \quad x \in (a,b), \tag{6.33}$$

$$u(a) = 0, \quad u(b) = 0, \tag{6.34}$$

where the function h models 1D source distribution in the interval $[a,b]$. Is it possible to describe a source concentrated at one point $x = \xi$ by a function h? The main difficulty consists of the zero measure (length) at one point in $[a,b]$. Let ε be a sufficiently small positive number. Let the interval $(\xi - \varepsilon, \xi + \varepsilon)$ absorbs the heat $Q > 0$ per unit time. Then, the heat density is given by

$$f(x) = \begin{cases} -\frac{Q}{2\varepsilon}, & \xi - \varepsilon < x < \xi + \varepsilon, \\ 0, & \text{otherwise.} \end{cases}$$

Therefore, the heat density is properly defined in the small segment of the length 2ε and

$$h(x) = \frac{f(x)}{\lambda} = \begin{cases} -\frac{Q}{2\varepsilon\lambda}, & \xi - \varepsilon < x < \xi + \varepsilon, \\ 0, & \text{otherwise.} \end{cases}$$

Is it possible to define the heat density concentrated at the point $x = \xi$? This question is equivalent to the following: how does one define the function h as ε tends to zero? A brave guy, not familiar with the generalized functions in the Sobolev spaces who is not afraid of anything including infinity, could propose the following

$$h(x) = \begin{cases} -\infty, & x = \xi, \\ 0, & x \neq \xi. \end{cases} \tag{6.35}$$

[2]Franklin D. Roosevelt, `brainyquote.com/quotes/quotes/f/franklind122780.html`

From the common point of view the infinity in (6.35) is quite natural, since touching a wire by hand at the sink point $x = \xi$ yields a conjecture that it is very (infinitely) cold at the point. From other side, instead of Q one can take $2Q$ and take $\varepsilon \to 0$. However, nobody sees this 2 at infinity[3].

Anyway, we will try to properly describe infinity in formula (6.35). First, we set aside the point $x = \xi$ and consider what is going on at the rest points. In Sec.6.5 we shall come back to the point $x = \xi$. Since $h(x) = 0$ for $x \neq \xi$, hence the function u satisfies the conditions

$$u''(x) = 0 \quad \text{for } x \in (a, \xi) \cup (\xi, b) \tag{6.36}$$

$$u(a) = 0, \quad u(b) = 0. \tag{6.37}$$

It follows from (6.36) that u is a linear function in (a, ξ) and it is another linear function in (ξ, b). Using (6.37) we get that u has the form

$$u(x) = \begin{cases} C_1(x - a), & a < x < \xi, \\ C_2(b - x), & \xi < x < b, \end{cases} \tag{6.38}$$

where C_1 and C_2 are undetermined constants.

We now pay attention to the limit values of the functions $u(x)$ and $u'(x)$ as x tends to ξ from the left and right sides. In the considered model, $u(x)$ describes the temperature distribution. The limit values

$$u(\xi - 0) := \lim_{x \to \xi, x < \xi} u(x), \quad \text{and} \quad u(\xi + 0) := \lim_{x \to \xi, x > \xi} u(x)$$

denote the temperatures at the point $x = \xi$ from different sides. If the considered wire has no defect at the point $x = \xi$ the temperatures from its different sides must coincide:

$$u(\xi - 0) = u(\xi + 0). \tag{6.39}$$

The value $-\lambda u'(\xi - 0)$ gives the total heat going into the point $x = \xi$ at the left side[4]. In the same time, the total heat outgoing from $x = \xi$ at the right side is equal to $-\lambda u'(\xi + 0)$. We have a sink at $x = \xi$. Let the intensity of this sink hold Q. This means that the heat Q disappears at the point $x = \xi$ per unit time. Therefore, the heat balance at $x = \xi$ has the form

$$q(\xi - 0) - q(\xi + 0) = -Q \quad \Leftrightarrow \quad u'(\xi - 0) - u'(\xi + 0) = \frac{Q}{\lambda}. \tag{6.40}$$

This implies that the derivative u' is not continuous at $x = \xi$ and its jump at this point holds $\frac{Q}{\lambda}$.

---

[3]The writer John Green: "Some infinities are bigger than other infinities".

[4]This value can be negative. Then actually it is the total heat outgoing from $x = \xi$.

Using (6.38) we rewrite (6.39)–(6.40) as the system of linear algebraic equations on C_1 and C_2

$$\begin{cases} C_1(\xi - a) = C_2(b - \xi), \\ C_1 + C_2 = \frac{Q}{\lambda}. \end{cases} \tag{6.41}$$

Its solution has the form

$$C_1 = \frac{Q}{\lambda}\frac{b-\xi}{b-a}, \quad C_2 = \frac{Q}{\lambda}\frac{\xi - a}{b-a}. \tag{6.42}$$

Substitution of (6.42) into (6.38) gives the temperature distribution

$$u(x) = \frac{Q}{\lambda}\begin{cases} \frac{1}{b-a}(x-a)(b-\xi), & a \le x \le \xi, \\ \frac{1}{b-a}(\xi - a)(b-x), & \xi \le x \le b. \end{cases} \tag{6.43}$$

Comparing (6.43) with (6.28) we conclude that u is Green's function multiplied by $\frac{Q}{\lambda}$. Therefore, Green's function $g(x, \xi)$ describes the temperature distribution when the sink of intensity $Q = \lambda$ is located at $x = \xi$. The functions u, u' i u'' are displayed in Fig.6.9 near $x = \xi$.

A similar case is considered in Sec.6.2 concerning Heaviside's function (6.9).

6.5 The δ–function

Though the temperature distribution and the heat flux have been rigorously described in the previous section one question stays vague: the second derivative of temperature is equal to infinity at $x = \xi$.

The concept of infinity excites mankind from the very beginning when a man begun to count $1, 2, 3, \ldots$. From a mathematical point of view, the notation *infinity* is clarified in the definition of the limit $n \to \infty$. This section is devoted to the tame infinity in order to understand and properly apply expression (6.35). The following optimistic principle will help us.

Principle by Dirac[5]**.** *If it is forbidden, but we want it very much, then it is allowed at the initial stage. We must carefully test it for further usage and find its justification.*

Our goal is to explain how generalized functions can be introduced in the most important example, the δ-function, known also as an impulse function.

[5]rediscovered later by Michail Zhvanetsky

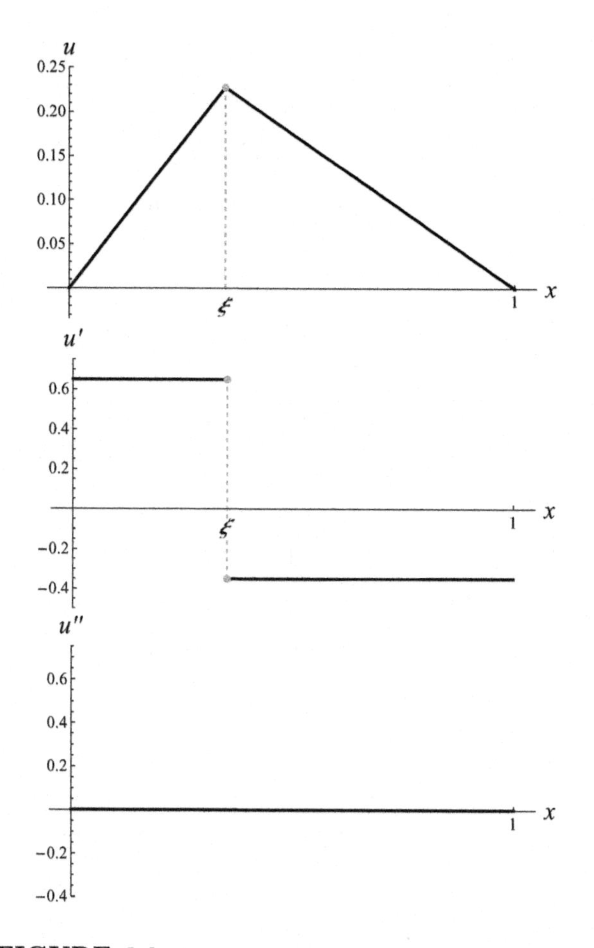

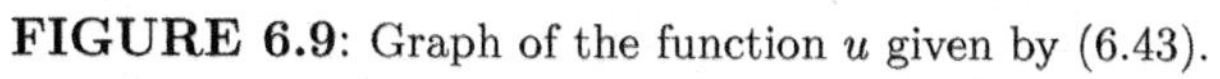
FIGURE 6.9: Graph of the function u given by (6.43).

Paul Dirac introduced generalized functions in 1927 and used them without necessary mathematical justifications. Later mathematicians strictly justified introduction of these functions as functionals in a space [21], [54] and as the limits of sequences [4]. Physicists were satisfied by this fact and confirmed that pure mathematics could sometimes be useful.

Consider a function $\Phi_1(x)$ which is positive on $\mathbb{R}$ and satisfies the condition

$$\int_{-\infty}^{+\infty} \Phi_1(x)\, dx = 1. \tag{6.44}$$

Hence, the area between the graph of $\Phi_1(x)$ and the x-axis holds 1. The functions $\frac{1}{\pi}\frac{1}{1+x^2}$ and $\frac{1}{\sqrt{\pi}}e^{-x^2}$ are examples of such functions. Introduce the functions

$$\Phi_m(x) = m\Phi_1(mx) \quad \text{for} \quad m = 2, 3, \ldots. \tag{6.45}$$

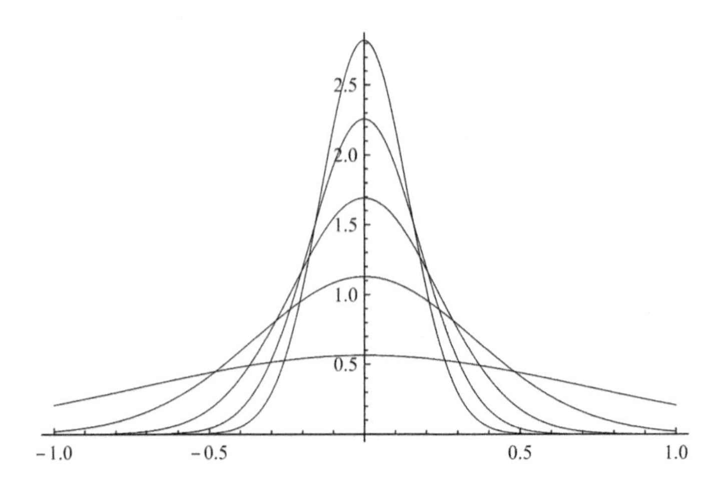

FIGURE 6.10: Graphs of $\Phi_m(x)$ for $m = 1, 2, \ldots, 5$ where $\Phi_1(x) = \frac{1}{\sqrt{\pi}}e^{-x^2}$.

Consider the limit $\lim_{m\to\infty} \Phi_m(x)$. It does not exist in a usual sense because

$$\lim_{m\to\infty} \Phi_m(x) = \begin{cases} +\infty, & x = 0, \\ 0, & x \neq 0. \end{cases} \tag{6.46}$$

But we want very much to have such a function. Hence, using the principle by Dirac we introduce the object $\delta(x)$ as the limit $\lim_{m\to\infty} \Phi_m(x)$. This generalized functions has the following properties:

a) $\delta(x) = 0$ **for any** $x \neq 0$;

b) $\delta(0) = +\infty$;

c) $\int_{-\infty}^{+\infty} \delta(x)\, dx = 1$.

Combining b) and c) we can see what double infinity means. Double infinity is equal to $2\delta(0)$ where $2\delta(x)$ is the limit of $2\Phi_m(x)$.

It follows from c) that

$$\int_{-\infty}^{+\infty} \delta(x)f(x)\,dx = f(0), \tag{6.47}$$

where $f(x)$ is a usual function continuous in a vicinity of zero. Formula (6.47) can be justified in the following way. The function $\delta(x)(f(x)-f(0))$ everywhere is equal to zero. Then, c) yields

$$\int_{-\infty}^{+\infty} \delta(x)(f(x)-f(0))\,dx = 0 \Leftrightarrow \int_{-\infty}^{+\infty} \delta(x)f(x)\,dx - f(0) = 0.$$

Actually, the following general formula takes place

$$\int_{-\infty}^{+\infty} \delta(x-c)f(x)\,dx = f(c). \tag{6.48}$$

Green's function $g(x,\xi)$ satisfies the problem

$$\frac{d^2u}{dx^2} = -\delta(x-\xi) \quad \text{for} \quad x \in (a,b), \tag{6.49}$$

$$u(a) = 0, \quad u(b) = 0. \tag{6.50}$$

This follows from (6.26) for $h(x) = \delta(x-\xi)$ and (6.48)

$$u(x) = \int_a^b g(x,\eta)\delta(\eta-\xi)\,d\xi = g(x,\xi).$$

Mathematica can work with the function $\delta(x)$ implemented as **DiracDelta**:

```
In[1]:= Integrate[DiracDelta[x - c] f[x], {x, -∞, ∞},
          Assumptions → c ∈ Reals]

Out[1]= f[c]

In[2]:= ∂x HeavisideTheta[x]

Out[2]= DiracDelta[x]

In[3]:= DiracDelta[1.2]

Out[3]= 0
```

Further reading. We recommend [55, 52].

Exercises

1. Prove Cauchy's formula (6.24) with *Mathematica*.

2. Investigate the functions $\frac{m}{\pi}\frac{1}{1+m^2x^2}$ for $m \to \infty$ using the operator **`Animate`**.

   For creating animation in MATLAB refer to Example Box 8.1 on page 175.

3. Calculate the integral

$$\int_{-\infty}^{+\infty}\int_{-\infty}^{+\infty}\int_{-\infty}^{+\infty} \delta(x)\delta(y)\delta(z)\ \mathrm{d}x\ \mathrm{d}y\ \mathrm{d}z.$$

Part III

Advanced Applications

Chapter 7

Vector analysis

7.1 Euclidean space $\mathbb{R}^3$ 133
7.1.1 Polar coordinates 135
7.1.2 Cylindrical coordinates 135
7.1.3 Spherical coordinates 137
7.2 Scalar, vector and mixed products 138
7.3 Rotation of bodies 142
7.4 Scalar, vector and mixed product in *Mathematica* 144
7.5 Tensors 145
7.6 Scalar and vector fields 148
7.6.1 Gradient 148
7.6.2 Divergence 152
7.6.3 Curl 156
7.6.4 Formulae for gradient, divergence and curl 156
7.7 Integral theorems 159
Exercises 161

This chapter can be considered as an introduction to vector analysis with applications in main to mechanics. Computer implementation of vector analysis is widely used.

7.1 Euclidean space $\mathbb{R}^3$

Following classic mechanics (Newtonian mechanics) the Euclidean space $\mathbb{R}^3$ is considered as a place of action and time t is an independent parameter. The notation $\mathbf{x}$ is used for the vector $\overrightarrow{OX}$ connecting O, the origin of $\mathbb{R}^3$, to a point X. The same notation[1] $\mathbf{x}$ is used for the point $X \in \mathbb{R}^3$. The standard orthogonal basis of $\mathbb{R}^3$ consists of the vectors

$$\mathbf{i}_1 = (1,0,0), \quad \mathbf{i}_2 = (0,1,0), \quad \mathbf{i}_3 = (0,0,1). \tag{7.1}$$

[1]There is a formal difference between points and vectors. Points are considered as geometric sets. Algebraic manipulations can be applied to vectors. For instance, a linear combination of vectors is correctly defined. But for points, it should be agreed upon vectors [20, p.11]. In the present book, such a difference between points and vectors is not essential.

Any vector $= \mathbf{x} \in \mathbb{R}^3$ can be presented uniquely in the form

$$\mathbf{x} = x_1\mathbf{i}_1 + x_2\mathbf{i}_2 + x_3\mathbf{i}_3 = (x_1, x_2, x_3). \tag{7.2}$$

The triple (x_1, x_2, x_3) denotes the Cartesian coordinate of the vector $\mathbf{x}$.

Physical laws have to be stated in the space $\mathbb{R}^3$. Frequently, it is easy to do using the fixed basis $(\mathbf{i}_1, \mathbf{i}_2, \mathbf{i}_3)$. However, physical laws must be invariant with respect to coordinates. A separate question exists in the form of physical laws independent on coordinates. This theoretical question is touched upon in Sec.7.5. In order to give constructive representations of the objects of vector analysis we investigate dependence of vector equations on the basis. Let the vectors $\mathbf{i}_1'$, $\mathbf{i}_2'$, $\mathbf{i}_3'$ form another orthogonal basis $(\mathbf{i}_1', \mathbf{i}_2', \mathbf{i}_3')$ with the same origin $\mathbf{0} = (0, 0, 0)$. Then, the basic vectors are related by formulae

$$\mathbf{i}_m' = \sum_{k=1,2,3} \alpha_{mk}\mathbf{i}_k, \tag{7.3}$$

where the coefficients α_{mk} form a matrix A. This matrix can be treated as a linear map of $\mathbb{R}^3$ onto $\mathbb{R}^3$. For shortness, let us omit the sign $\sum$. Then, formula (7.3) becomes

$$\mathbf{i}_m' = \alpha_{mk}\mathbf{i}_k. \tag{7.4}$$

Such a notation is called the Einstein summation convention where the sum is performed on the subscript variable k appearing twice in (7.4).

Example 7.1. Let $\mathbf{i}_3' = \mathbf{i}_3$ in (7.3) and the basis $(\mathbf{i}_1, \mathbf{i}_2)$ is transformed into the basis $(\mathbf{i}_1', \mathbf{i}_2')$, i.e., actually only the plane $\mathbb{R}^2$ is transformed. Then the matrix A becomes

$$A = \begin{pmatrix} \cos\theta & \sin\theta & 0 \\ -\sin\theta & \cos\theta & 0 \\ 0 & 0 & 1 \end{pmatrix}, \tag{7.5}$$

where the vectors $\mathbf{i}_1'$ and $\mathbf{i}_1$ form the angle θ. The coefficients $\cos\theta$ and $\sin\theta = \cos(\frac{\pi}{2} - \theta)$ are called *direction cosines.*

In a general case, the matrix A has a similar structure

$$A = \begin{pmatrix} \alpha_{11} & \alpha_{12} & \alpha_{13} \\ \alpha_{21} & \alpha_{22} & \alpha_{23} \\ \alpha_{31} & \alpha_{32} & \alpha_{33} \end{pmatrix}, \tag{7.6}$$

where the coefficient α_{mk} is the direction cosine of the angle between the vectors $\mathbf{i}_m'$ and $\mathbf{i}_k$.

7.1.1 Polar coordinates

Polar coordinate system is introduced in the plane $\mathbb{R}^2$. Any point $M \in \mathbb{R}^2$ possesses two coordinates r and θ, where r is equal to the distance $|OM|$ and θ is the angle formed by the vector $\overrightarrow{OM}$ and the x-axis (for $M \neq O$).

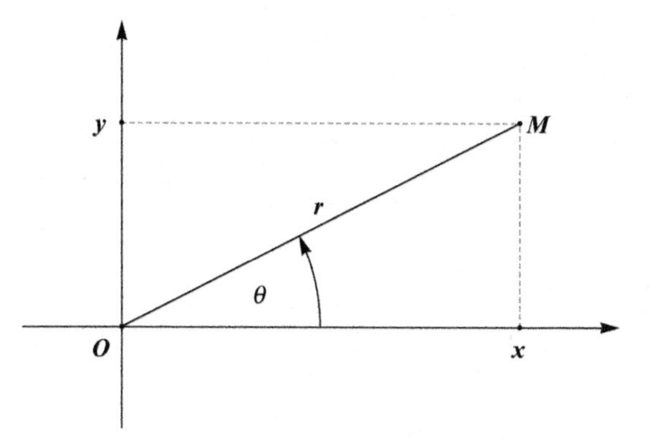

FIGURE 7.1: Polar coordinate system.

Let a point M have the Cartesian coordinate (x, y). Then, the polar coordinates (r, θ) are calculated by formulae

$$r = \sqrt{x^2 + y^2}, \quad \text{tg}\,\theta = \frac{y}{x}. \tag{7.7}$$

Conversely

$$x = r\cos\theta, \quad y = r\sin\theta. \tag{7.8}$$

It is worth noting that formula

$$\theta = \text{arctg}\,\frac{y}{x}. \tag{7.9}$$

is valid only for $x > 0$. In the case $x < 0$,

$$\theta = \text{arctg}\,\frac{y}{x} + \pi. \tag{7.10}$$

$\theta = \frac{\pi}{2}$ for $x = 0$ and $y > 0$; $\theta = -\frac{\pi}{2}$ for $x = 0$ and $y < 0$.

7.1.2 Cylindrical coordinates

Cylindrical coordinates are introduced in $\mathbb{R}^3$ similar to the polar coordinates. Let a point $M = (x, y, z)$ be given in the Cartesian coordinates. Let r denote the distance between the point M and the z-axis; ϕ the angle between the x-axis and the plane which contains the z-axis and the point M (see Fig.7.2). The triple (r, ϕ, z) is called the cylindrical coordinates of the point M. One can see that

$$r = \sqrt{x^2 + y^2}, \tag{7.11}$$

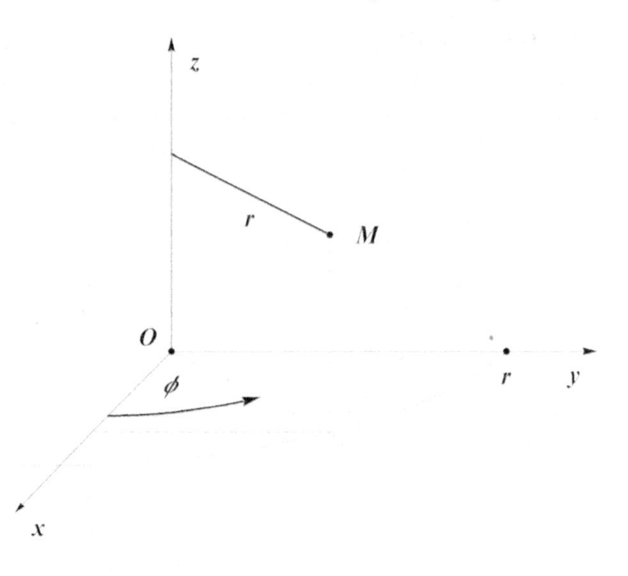

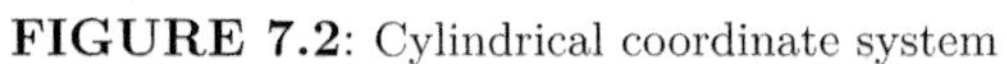

FIGURE 7.2: Cylindrical coordinate system

and

$$\sin\phi = \frac{y}{\sqrt{x^2+y^2}}, \qquad \cos\phi = \frac{x}{\sqrt{x^2+y^2}}. \tag{7.12}$$

The Cartesian coordinates of the point M are expressed through the cylindrical coordinates by formulae

$$x = r\cos\theta, \quad y = r\sin\theta. \tag{7.13}$$

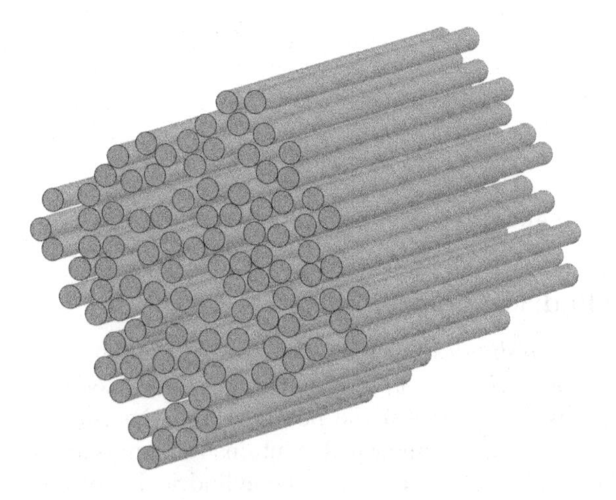

FIGURE 7.3: Fibrous composite.

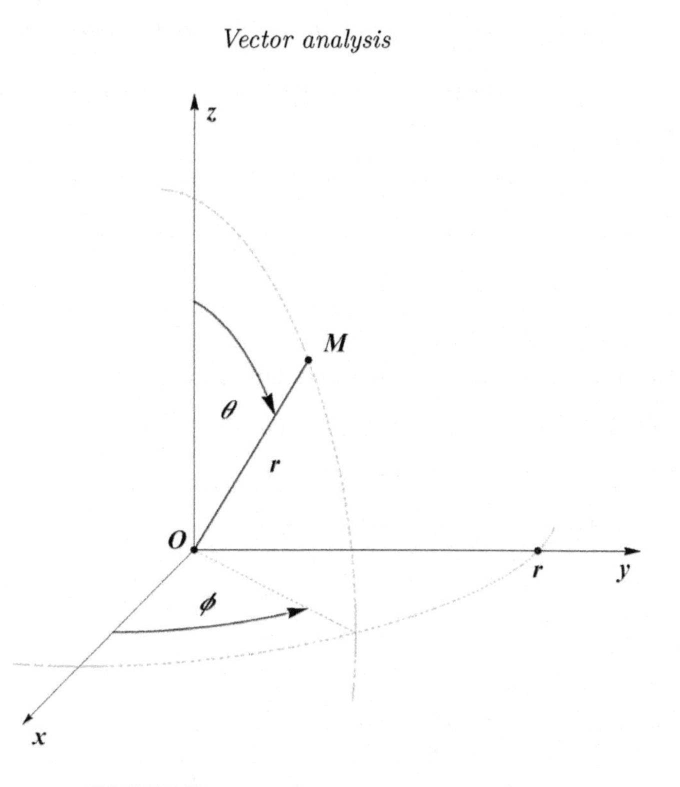

FIGURE 7.4: Spherical coordinate system.

Polar and cylindrical coordinates are convenient for the description of fibrous composites when unidirectional infinite cylinders are embedded in host material as shown in Fig.7.3. A local field can be decomposed onto two fields, perpendicular and parallel to cylinders. The resulting fields can be independently investigated in polar and cylindrical coordinates.

7.1.3 Spherical coordinates

Consider a point $M = (x, y, z)$ in the Cartesian coordinates. Let O denote the origin and r the distance between the points M and O. Let θ be the angle between the z-axis and the vector OM (for $M \neq O$); ϕ the angle between the x-axis and the plane which contains the z-axis and the point M (see Fig.7.4). The triple (r, θ, ϕ) is called the spherical coordinates of the point M. The angles θ and ϕ are related to the Cartesian coordinates by equations

$$\cos\theta = \frac{z}{r}, \quad \sin\theta = \frac{\sqrt{x^2 + y^2}}{r} \tag{7.14}$$

and

$$\sin\phi = \frac{y}{\sqrt{x^2 + y^2}}, \quad \cos\phi = \frac{x}{\sqrt{x^2 + y^2}}. \tag{7.15}$$

The Cartesian coordinates are expressed through the spherical coordinates by formulae

$$x = r\sin\theta\cos\phi, \quad y = r\sin\theta\sin\phi, \quad z = r\cos\theta. \tag{7.16}$$

7.2 Scalar, vector and mixed products

Definition 7.1. Scalar product of two vectors $\mathbf{a}$ and $\mathbf{b}$ is defined by formula

$$\mathbf{a}\cdot\mathbf{b} = |\mathbf{a}||\mathbf{b}|\cos(\mathbf{a},\mathbf{b}), \tag{7.17}$$

where $\theta = (\mathbf{a},\mathbf{b})$ is the angle between $\mathbf{a}$ and $\mathbf{b}$.

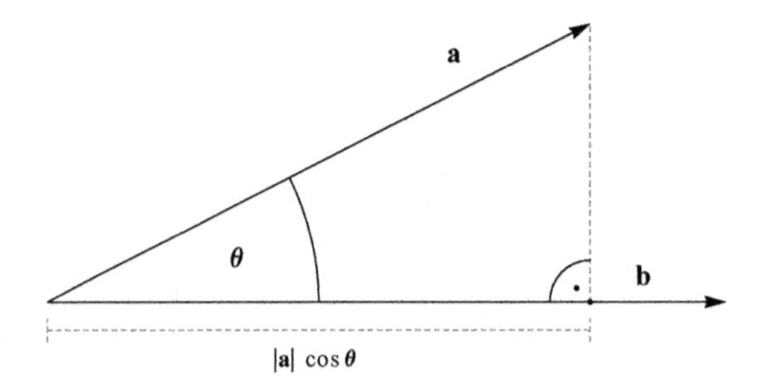

FIGURE 7.5: Geometrical interpretation of the scalar product.

The scalar product in the Cartesian coordinates is calculated by formula

$$\mathbf{a}\cdot\mathbf{b} = a_1b_1 + a_2b_2 + a_3b_3 = a_ib_i, \tag{7.18}$$

where $\mathbf{a} = (a_1, a_2, a_3)$ and $\mathbf{b} = (b_1, b_2, b_3)$.

Example 7.2. Consider the straight motion of a material point from a point A to a point B. Let a force $\mathbf{F}$ act on the material point at every point $\mathbf{x}$ of the interval AB.

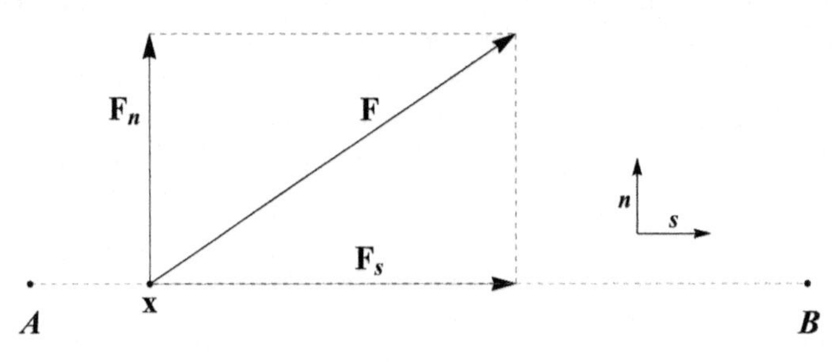

FIGURE 7.6: Force $\mathbf{F}$ and its decomposition.

Introduce unit vectors $\mathbf{s}$ directed along AB and $\mathbf{n}$ perpendicular to AB as shown in Fig.7.6. The force $\mathbf{F}$ can be presented as the sum $\mathbf{F} = \mathbf{F}_s + \mathbf{F}_n$, where the scalar product $\mathbf{F}_s = \mathbf{F} \cdot \mathbf{s}$ is the projection of $\mathbf{F}$ on the line AB; the scalar product $\mathbf{F}_n = \mathbf{F} \cdot \mathbf{n}$ is the projection onto a line perpendicular to AB. The force component $\mathbf{F}_n$ does not impact onto the motion of the material point along the interval AB, but the vector $\mathbf{F}_s$ does.

Definition 7.2. The vector $\mathbf{c} := \mathbf{a} \times \mathbf{b}$ is called the vector product of two vectors $\mathbf{a}$ and $\mathbf{b}$ if

i) $|\mathbf{c}| = |\mathbf{a}||\mathbf{b}| \sin(\mathbf{a}, \mathbf{b})$,

ii) $\mathbf{c} \perp \mathbf{a}$ and $\mathbf{c} \perp \mathbf{b}$,

iii) the direction of the resulting vector $\mathbf{c}$ is uniquely determined by the mutual location of $\mathbf{a}$ and $\mathbf{b}$ explained in Fig.7.7.

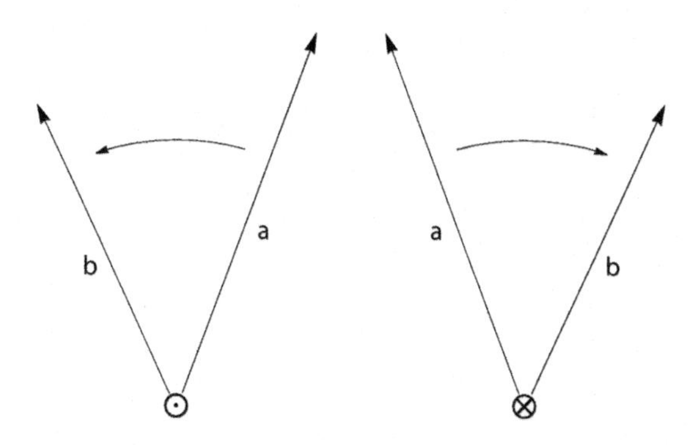

FIGURE 7.7: To item iii) of Definition 7.2: $\odot$ means that the vector $\mathbf{c} = \mathbf{a} \times \mathbf{b}$ is directed to the Reader; $\otimes$ from the Reader.

Remark 7.1. If the vectors $\mathbf{a}$ and $\mathbf{b}$ are parallel, then $\sin(\mathbf{a}, \mathbf{b}) = 0$ and i) implies that $\mathbf{c} = \mathbf{0}$.

The definition of the angle $(\mathbf{a}, \mathbf{b})$ supposes that the angle begins from the vector $\mathbf{a}$ and ends on the vector $\mathbf{b}$, hence it has the sign. It does not matter for the scalar product since the angle $(\mathbf{a}, \mathbf{b})$ is in (7.17) under the even function cosine.

The area of the parallelogram formed by the vectors $\mathbf{a}$ and $\mathbf{b}$ holds $|\mathbf{c}|$. It follows from Definition 7.2 that the swap of $\mathbf{a}$ and $\mathbf{b}$ changes the sign of the vector product, i.e.,

$$\mathbf{a} \times \mathbf{b} = -\mathbf{b} \times \mathbf{a}. \tag{7.19}$$

The basis vectors $\mathbf{i}_k$ satisfy equations

$$\mathbf{i}_k \times \mathbf{i}_m = \delta_{km}, \tag{7.20}$$

where δ_{km} stands for the Kronecker symbol.

The vector product can be calculated in terms of the Cartesian coordinates of the vectors $\mathbf{a}$ and $\mathbf{b}$

$$\mathbf{a} \times \mathbf{b} = \begin{vmatrix} \mathbf{i}_1 & \mathbf{i}_2 & \mathbf{i}_3 \\ a_1 & a_2 & a_3 \\ b_1 & b_2 & b_3 \end{vmatrix} = (a_2 b_3 - a_3 b_2, a_3 b_1 - a_1 b_3, a_1 b_2 - a_2 b_1). \quad (7.21)$$

Example 7.3. The *moment of a force* about a point is defined as follows. Let a beam of the length ℓ be fixed at the point O. Let a mass m be located at the other end N of the beam. Then the gravitational force is calculated by formula $F = mg$. The moment of F about O is defined as

$$M = F\ell. \quad (7.22)$$

This is a preliminary definition which does not take into account that force is expressed by a vector. The moment of force about a point has the following vector interpretation. Let $\mathbf{x}$ denote the vector $\overrightarrow{ON}$ and $\mathbf{F}$ the force. Then, the moment is introduced as the vector

$$\mathbf{M} = \mathbf{x} \times \mathbf{F}. \quad (7.23)$$

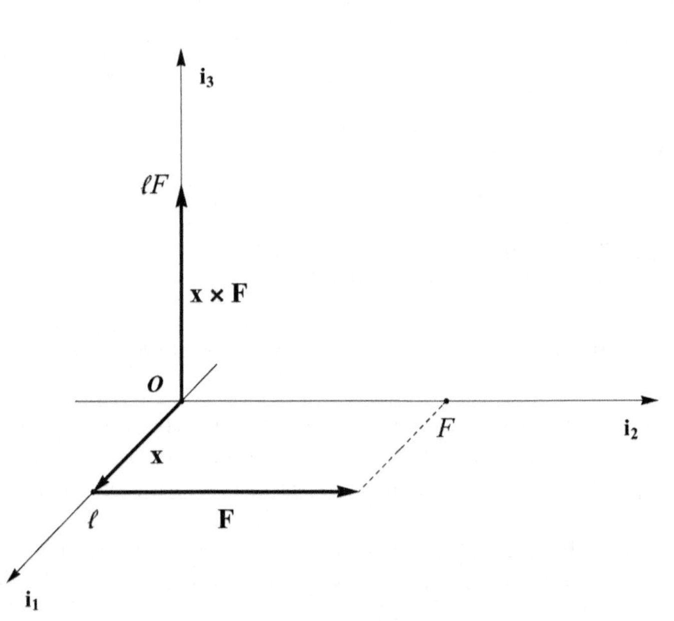

FIGURE 7.8: Geometric interpretation of formula (7.23).

Let the Cartesian coordinates be chosen in such a way that $\mathbf{x} = (\ell, 0, 0)$ and $\mathbf{F} = (0, F, 0)$ (see Fig.7.8). Formulae (7.21) and (7.23) yield $\mathbf{M} = \ell F \mathbf{i}_3$. Hence, $|\mathbf{M}| = M$. The assignment of the direction to the moment, i.e., its presentation in the vector form (7.23) has advantages in classical mechanics (see Example 7.6 on page 156).

Definition 7.3. The scalar value

$$V = (\mathbf{a} \times \mathbf{b}) \cdot \mathbf{c}. \tag{7.24}$$

is called the mixed product of the vectors $\mathbf{a}$, $\mathbf{b}$ and $\mathbf{c}$.

According to (7.21) and (7.18) the mixed product cam be calculated by the formula

$$(\mathbf{a} \times \mathbf{b}) \cdot \mathbf{c} = \begin{vmatrix} c_1 & c_2 & c_3 \\ a_1 & a_2 & a_3 \\ b_1 & b_2 & b_3 \end{vmatrix}. \tag{7.25}$$

It follows from (7.25) and the properties of determinants that the vectors $\mathbf{a}$, $\mathbf{b}$ and $\mathbf{c}$ are linearly independent if and only if $(\mathbf{a} \times \mathbf{b}) \cdot \mathbf{c} \neq 0$. Moreover, the following formula holds

$$(\mathbf{a} \times \mathbf{b}) \cdot \mathbf{c} = (\mathbf{b} \times \mathbf{c}) \cdot \mathbf{a} = (\mathbf{c} \times \mathbf{a}) \cdot \mathbf{b}. \tag{7.26}$$

It follows from (7.21) that

$$(\mathbf{a} + \mathbf{b}) \times \mathbf{c} = \mathbf{a} \times \mathbf{c} + \mathbf{b} \times \mathbf{c}. \tag{7.27}$$

The following formulae are also fulfilled

$$\mathbf{a} \times (\mathbf{b} \times \mathbf{c}) = \mathbf{b}(\mathbf{a} \cdot \mathbf{c}) - \mathbf{c}(\mathbf{a} \cdot \mathbf{b}), \tag{7.28}$$

$$(\mathbf{a} \times \mathbf{b}) \times \mathbf{c} = \mathbf{b}(\mathbf{a} \cdot \mathbf{c}) - \mathbf{a}(\mathbf{b} \cdot \mathbf{c}), \tag{7.29}$$

$$(\mathbf{a} \times \mathbf{b}) \cdot (\mathbf{c} \times \mathbf{d}) = \begin{vmatrix} \mathbf{a} \cdot \mathbf{c} & \mathbf{a} \cdot \mathbf{d} \\ \mathbf{b} \cdot \mathbf{c} & \mathbf{b} \cdot \mathbf{d} \end{vmatrix}, \tag{7.30}$$

$$\mathbf{a} \times (\mathbf{b} \times \mathbf{c}) + \mathbf{b} \times (\mathbf{c} \times \mathbf{a}) + \mathbf{c} \times (\mathbf{a} \times \mathbf{b}) = \mathbf{0}, \tag{7.31}$$

$$(\mathbf{a} \times \mathbf{b}) \cdot (\mathbf{c} \times \mathbf{d}) + (\mathbf{b} \times \mathbf{c}) \cdot (\mathbf{a} \times \mathbf{d}) + (\mathbf{c} \times \mathbf{a}) \cdot (\mathbf{b} \times \mathbf{d}) = \mathbf{0}. \tag{7.32}$$

The short proof of formulae (7.29)-(7.32) is based on (7.28) and outlined below. Formula (7.29) follows from (7.28) after the cyclic substitution $\mathbf{b} \to \mathbf{a} \to \mathbf{c} \to \mathbf{b}$ and using (7.19). Formula (7.30) is obtained from (7.26) written as $(\mathbf{a} \times \mathbf{b}) \cdot \mathbf{e} = (\mathbf{b} \times \mathbf{e}) \cdot \mathbf{a}$ with $\mathbf{e} = \mathbf{c} \times \mathbf{d}$ and by use of (7.28). Formula (7.31) is deduced from (7.28) for corresponding triples with further their addition. Formula (7.32) is proved on the basis of (7.30) by use of similar arguments.

The demonstration of (7.28) requires direct and long manipulations (see for instance [13], p. 22). A way to avoid this hard work is to use symbolic computations. We shall do it in Sec.7.4.

Example 7.4. Consider a tetrahedron shown in Fig.7.9 with faces T_i ($i = 1, 2, 3, 4$). Let $\mathbf{S}_i$ be a vector perpendicular to T_i directed out of the tetrahedron and $|\mathbf{S}_i|$ be equal to the area of T_i. Then,

$$\mathbf{S}_1 + \mathbf{S}_2 + \mathbf{S}_3 + \mathbf{S}_4 = \mathbf{0}. \tag{7.33}$$

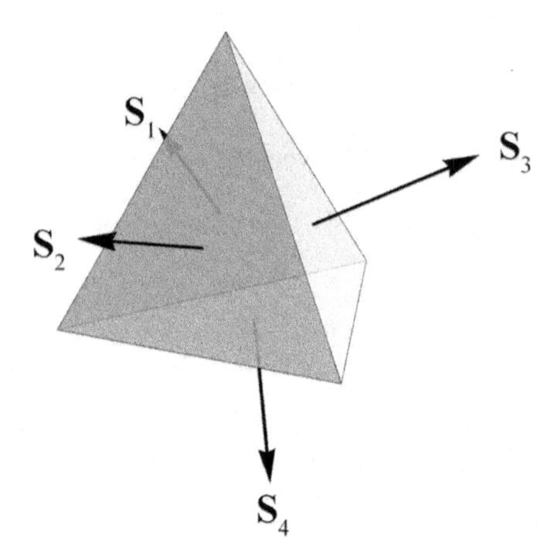

FIGURE 7.9: Tetrahedron and the vectors $\mathbf{S}_i$.

First proof. Any tetrahedron can be formed by three vectors $\mathbf{a}$, $\mathbf{b}$ and $\mathbf{c}$ having the same tail. These three vectors can be considered as edges of the tetrahedron. The other three edges are associated with the vectors $\mathbf{b}-\mathbf{a}$, $\mathbf{c}-\mathbf{a}$ and $\mathbf{c}-\mathbf{b}$. Every vector $\mathbf{S}_i$ is expressed through the vector products

$$\mathbf{S}_1 = \mathbf{b} \times \mathbf{a}, \quad \mathbf{S}_2 = \mathbf{c} \times \mathbf{b}, \quad \mathbf{S}_3 = \mathbf{a} \times \mathbf{c}, \quad \mathbf{S}_4 = (\mathbf{b}-\mathbf{a}) \times (\mathbf{c}-\mathbf{a}). \tag{7.34}$$

Calculation of $\mathbf{S}_1 + \mathbf{S}_2 + \mathbf{S}_3 + \mathbf{S}_4$ yields (7.33).

Second proof. Let a balloon having the tetrahedron shape be filled by a gas. Let its weight be at the state of equilibrium with the gravitational force. Hence, it is at rest in the air. The gas interior of the balloon has the pressure p. The force acting on each face is equal to $p\mathbf{S}_i$. The resultant force $\mathbf{F}$ must vanish, because if $\mathbf{F}$ does not vanish the balloon should move in the direction $\mathbf{F}$ without any exterior reason. This yields equality (7.33).

The second proof can be easily extended to an arbitrary polyhedra. Many physical proofs of mathematical theorems are selected in the excellent book by Levi [38].

7.3 Rotation of bodies

Let a material point located at a point M rotate about the x_3-axis in a counterclockwise direction. Introduce the vector $\mathbf{x} = \overrightarrow{OM}$ perpendicular to the x_3-axis. One can assume that O is the origin. Let $\boldsymbol{\omega}$ be the angular velocity measured, for instance, in degrees per second (see Fig.7.10). The vector $\boldsymbol{\omega}$ is directed to the x_3-axis.

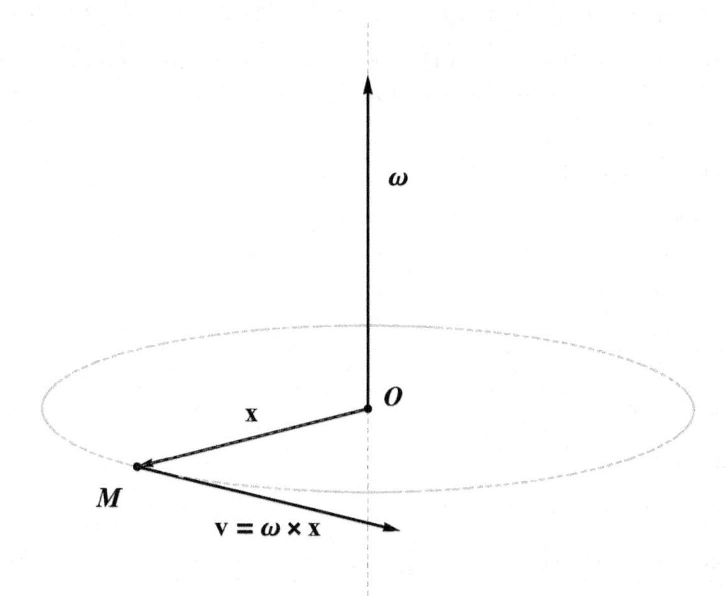

FIGURE 7.10: Rotation of the point M about the x_3-axis with the angle velocity $\boldsymbol{\omega}$.

The line velocity $\mathbf{v}$ of the point M with the coordinate $\mathbf{x}$ is related to the *angle velocity* $\boldsymbol{\omega}$ by formula

$$\mathbf{v} = \boldsymbol{\omega} \times \mathbf{x}. \tag{7.35}$$

This simple mathematical formula has evident mechanical interpretation that the angle velocities can be added, i.e.,

$$(\boldsymbol{\omega}_1 + \boldsymbol{\omega}_2) \times \mathbf{x} = \boldsymbol{\omega}_1 \times \mathbf{x} + \boldsymbol{\omega}_2 \times \mathbf{x}. \tag{7.36}$$

Consider now a rotation of a material point of mass m located at $\mathbf{x}$ at time t. *The angular momentum* of the material point about O is introduced as

$$\mathbf{L} = m\mathbf{x} \times \mathbf{v}. \tag{7.37}$$

The kinetic energy of the material point is calculated on the basis of formulae (7.35) and (7.37)

$$K = \frac{m|\mathbf{v}|^2}{2} = \frac{m|\boldsymbol{\omega}|^2|\mathbf{x}|^2}{2} = \frac{1}{2}|\mathbf{L}||\boldsymbol{\omega}|. \tag{7.38}$$

7.4 Scalar, vector and mixed product in *Mathematica*

Manipulations described in the previous sections can be easily performed with *Mathematica*. The operator **Dot** responds to the scalar product. It can be used in various forms:

```
In[1]:= Dot[{x, y, z}, {u, v, w}]

Out[1]= u x + v y + w z
```

or

```
In[2]:= {x, y, z}.{u, v, w}

Out[2]= u x + v y + w z
```

The operator **Cross** carries out the vector product:

```
In[3]:= Cross[{x, y, z}, {u, v, w}]

Out[3]= {w y - v z, -w x + u z, v x - u y}
```

One can also use the sequence of buttons *Esc*, to type "cross" and *Esc* again to get the vector product:

```
In[4]:= {x, y, z}×{u, v, w}

Out[4]= {w y - v z, -w x + u z, v x - u y}
```

In order to prove formula (7.28) with *Mathematica* we introduce the triple of vectors $\mathbf{a}, \mathbf{b}, \mathbf{c}$ by coordinates:

```
In[5]:= a = {a1, a2, a3};
        b = {b1, b2, b3};
        c = {c1, c2, c3};
```

Calculate the left side L of (7.28), i.e., $\mathbf{a} \times (\mathbf{b} \times \mathbf{c})$:

```
In[8]:= L = a× (b×c)

Out[8]= {-a2 b2 c1 - a3 b3 c1 + a2 b1 c2 + a3 b1 c3,
         a1 b2 c1 - a1 b1 c2 - a3 b3 c2 + a3 b2 c3,
         a1 b3 c1 + a2 b3 c2 - a1 b1 c3 - a2 b2 c3}
```

Now, calculate the right side P of (7.28), i.e., $\mathbf{b}(\mathbf{a} \cdot \mathbf{c}) - \mathbf{c}(\mathbf{a} \cdot \mathbf{b})$:

```
In[9]:= P = b (a.c) - c (a.b)

Out[9]= {- (a1 b1 + a2 b2 + a3 b3) c1 + b1 (a1 c1 + a2 c2 + a3 c3),
         - (a1 b1 + a2 b2 + a3 b3) c2 + b2 (a1 c1 + a2 c2 + a3 c3),
         - (a1 b1 + a2 b2 + a3 b3) c3 + b3 (a1 c1 + a2 c2 + a3 c3)}
```

Simplify the expression P:

```
In[10]:= P = % // Simplify

Out[10]= {-a2 b2 c1 - a3 b3 c1 + a2 b1 c2 + a3 b1 c3,
          a1 b2 c1 - a1 b1 c2 - a3 b3 c2 + a3 b2 c3,
          a1 b3 c1 + a2 b3 c2 - a1 b1 c3 - a2 b2 c3}
```

Let us check that the obtained expressions are equal, i.e., that both sides of (7.28) coincide. It can be done by two methods. First, both sides can be compared by use of the double equality **==** [2]. *Mathematica* gives the logical value of this comparison:

```
In[11]:= L == P

Out[11]= True
```

Another verification of $L == P$ is based on subtraction[3].

```
In[12]:= L - P

Out[12]= {0, 0, 0}
```

7.5 Tensors

The tensor theory is used in transformations of equations when coordinate systems change. Any scalar value can be considered as a *tensor of order* 0. For instance, the temperature distribution $u(\mathbf{x})$ is an example of tensor of order 0. A scalar does not change with a change of coordinates: $u(\mathbf{x}) = u[f(\mathbf{y})]$, where $\mathbf{x} = f(\mathbf{y})$.

Vectors are not invariant under changes of coordinates. For instance, the vector $\mathbf{a} = (1, 2, 3)$ written in the basis (7.1) has other coordinates in another basis. Let the vector $\mathbf{a}$ expresses a force. This physical vector is the same physical object which does not depend on the basis. It is expressed in different mathematical forms in different coordinates. This situation is similar to representations of five fingers on a hand. It can be presented as 5 in the decimal numbering system and as 101 in the binary system. But the number five is an absolute reality which does not depend on the form of representation.

Let A denote a rotation matrix in $\mathbb{R}^3$ about the origin. Let it be written in the form of the direction cosines (7.6). Then, a vector written in terms of

[2]The single equality "**=**" serves the definition to an expression.

[3]The operator Simplify could be added in such cases since zero could not be explicitly seen.

the coordinates a_k after the transformation A takes new coordinates a'_m in accordance with the rule (cf. (7.4))

$$a'_m = \alpha_{mk} a_k, \tag{7.39}$$

where α_{mk} are the direction cosines. The latter equation can be written in the vector-matrix form

$$\mathbf{a}' = A\mathbf{a}. \tag{7.40}$$

Any physical vector is a *tensor of order* 1. In other words, a set of all vectors related by (7.39) or (7.40) is a tensor of order 1.

The next *tensor of order* 2 is associated to matrices. Formally, a tensor of order 2 is introduced as a set of matrices related by special formulae under rotations. Let a matrix B express a physical value in a coordinate system, B' be the same physical value in another coordinates related to the original one by a matrix $A = \{\alpha_{km}\}$. Here, $\alpha_{km} = \cos(\mathbf{i}'_k, \mathbf{i}_m)$ are the direction cosines, $\mathbf{i}_m$ and $\mathbf{i}'_k$ are the basis vectors of the original and transformed coordinate systems. The matrix B is a tensor of order 2 if for any A the elements of the matrices B and B' are related by formula

$$b'_{jk} = \alpha_{jl}\alpha_{mk} b_{lm}, \tag{7.41}$$

where the sum runs over l and m.

In the present section, we discuss the inertia tensors $\mathbf{L} = (L_1, L_2, L_3)$ given by (7.37). Substitute (7.35) into (7.37)

$$\mathbf{L} = m\mathbf{x} \times (\boldsymbol{\omega} \times \mathbf{x}). \tag{7.42}$$

Formula (7.28) implies that

$$\mathbf{L} = m[\boldsymbol{\omega}(\mathbf{x} \cdot \mathbf{x}) - \mathbf{x}(\boldsymbol{\omega} \cdot \mathbf{x})]. \tag{7.43}$$

Let $\boldsymbol{\omega} = (\omega_1, \omega_2, \omega_3)$. Then, equation (7.43) can be expanded in the coordinate form

$$L_j = m(\omega_j x_l x_l - x_j \omega_k x_k), \tag{7.44}$$

where the sum is calculated on l and k. Let ω_j be written as the sum $\omega_j = \delta_{jk}\omega_k$. Then, (7.44) becomes

$$L_j = m I_{jk} \omega_k, \tag{7.45}$$

where

$$I_{jk} = m(\delta_{jk} x_l x_l - x_j x_k). \tag{7.46}$$

Calculate I_{jk} for $j, k = 1, 2, 3$

$$\begin{aligned}
I_{11} &= m(x_l x_l - x_1^2) = m(x_2^2 + x_3^2), \\
I_{22} &= m(x_1^2 + x_3^2), \\
I_{33} &= m(x_1^2 + x_2^2), \\
I_{12} &= I_{21} = -m x_1 x_2, \\
I_{13} &= I_{31} = -m x_1 x_3, \\
I_{23} &= I_{32} = -m x_2 x_3.
\end{aligned} \tag{7.47}$$

The elements I_{jk} constitute a matrix which represents the inertia tensor $\mathbf{I}$. It follows from (7.45) that

$$\mathbf{L} = m\mathbf{I} \cdot \boldsymbol{\omega}. \tag{7.48}$$

This means that the moment vector $\mathbf{L}$ (tensor of order 1) is equal to the product of $\mathbf{I}$ (tensor of order 2) and the vector $\boldsymbol{\omega}$ (tensor of order 1).

Consider n particles of masses m_i located at the points $\mathbf{x}_i$ $(i = 1, 2, \dots, n)$. The inertia tensor of this structure in the Cartesian coordinates has the form

$$\begin{aligned}
I_{11} &= \sum_{i=1}^{n} m_i(x_{i2}^2 + x_{i3}^2), \\
I_{22} &= \sum_{i=1}^{n} m_i(x_{i1}^2 + x_{i3}^2), \\
I_{33} &= \sum_{i=1}^{n} m_i(x_{i1}^2 + x_{i2}^2), \\
I_{12} &= I_{21} = -\sum_{i=1}^{n} m_i x_{i1} x_{i2}, \\
I_{13} &= I_{31} = -\sum_{i=1}^{n} m_i x_{i1} x_{i3}, \\
I_{23} &= I_{32} = -\sum_{i=1}^{n} m_i x_{i2} x_{i3},
\end{aligned} \tag{7.49}$$

where $\mathbf{x}_i = (x_{i1}, x_{i2}, x_{i3})$.

Consider now a solid body occupying a domain D of the density $\rho(\mathbf{x})$ dependent on $\mathbf{x}$. Applying the principle of transition *continuous* $\leftrightarrow$ *discrete* we calculate the inertia moment of this body. Consider, for instance, the tensor element I_{11}. Instead of the continuous body consider its small parts with the gravitational centers at the points $\mathbf{x}_i$ of masses m_i. Using the first formula (7.49) we obtain I_{11} for a discrete body. The mass m_i is represented in the form $m_i = \rho(\mathbf{x}_i)\Delta V_i$, where ΔV_i is the volume of the ith part. Passing to the limit $\max_i(\Delta V_i) \to 0$, in the first formula (7.49) we obtain the following triple integral over the domain D[4]

$$I_{11} = \int_D \rho(\mathbf{x})(x_2^2 + x_3^2)\, d\mathbf{x}, \tag{7.50}$$

where $\mathbf{x} = (x_1, x_2, x_3)$, $d\mathbf{x} = dx_1 dx_2 dx_3$.

Symmetric tensors of order 2 in the form of symmetric matrices frequently occur in applications. It follows from the matrix theory that any symmetric matrix is similar to a diagonal matrix with real elements. This means that

[4]Sometimes, the symbol $\int\int\int_D$ is used for 3D integrals.

there exists such a coordinate system in which the tensor is represented by means of a diagonal matrix.

The most important tensors of order 2 are the stress and deformation tensors in the elasticity theory [13].

7.6 Scalar and vector fields

A function of three variables $u : D \to \mathbb{R}$ defined in a domain $D \subset \mathbb{R}^3$ is called a *scalar field.* For instance, the temperature distribution $T = T(\mathbf{x})$ or the pressure $p = p(\mathbf{x})$ are scalar fields when $\mathbf{x} \in D$.

A *vector field* is introduced as a vector-function $\mathbf{a} : D \to \mathbb{R}^3$ defined in the set D. For instance, a force in D, i.e., a force given in the vector form at each point of D can be treated as a vector field. In the Cartesian coordinates, a vector field has the form $\mathbf{a}(\mathbf{x}) = (a_1(\mathbf{x}), a_2(\mathbf{x}), a_3(\mathbf{x}))$. Similar notations are used for tensor fields.

7.6.1 Gradient

Let a function u be differentiable in the domain D, i.e., let u possess partial derivatives continuous in D. The scalar field u yields the following vector field given in the Cartesian coordinates[5]

$$\text{grad } u \equiv \nabla u := \left(\frac{\partial u}{\partial x_1}, \frac{\partial u}{\partial x_2}, \frac{\partial u}{\partial x_3} \right). \tag{7.51}$$

The vector field (7.51) is called the *gradient* of u. Introduce the differential operator

$$\nabla := \left(\frac{\partial}{\partial x_1}, \frac{\partial}{\partial x_2}, \frac{\partial}{\partial x_3} \right). \tag{7.52}$$

Let $\mathbf{w}$ be a unit length vector. The gradient and the directional derivatives are discussed in calculus. They are related by equation

$$\frac{\partial u}{\partial \mathbf{w}} = \nabla u \cdot \mathbf{w}. \tag{7.53}$$

In the Cartesian coordinates, (7.53) becomes

$$\frac{\partial u}{\partial \mathbf{w}} = \frac{\partial u}{\partial x_1} w_1 + \frac{\partial u}{\partial x_2} w_2 + \frac{\partial u}{\partial x_3} w_3, \tag{7.54}$$

where $\mathbf{w} = (w_1, w_2, w_3)$.

A curve L can be considered as a set of points in $\mathbb{R}^3$. This geometric object

---

[5]the Geek letter ∇ pronounced as "nabla" is used.

is called the *support of the curve* L. In calculus, a curve is introduced as a triple of functions $P = \{x_1 = x_1(\tau), x_2 = x_2(\tau), x_3 = x_3(\tau)\}$ determined in an interval $[\tau_1, \tau_2]$; the triple P is called the *parametrization of the curve* L; τ is a parameter of L. For simplicity, we consider smooth curves when the functions $x_j(\tau)$ are continuously differentiable and

$$\sum_{j=1,2,3} [x_j'(\tau)]^2 \neq 0 \quad \text{in } (\tau_1, \tau_2).$$

Moreover, $x_j(\tau)$ $(j = 1, 2, 3)$ are continuous in the closed interval $[\tau_1, \tau_2]$. For example, the unit circle in the (x_1, x_2)-plane with the center at the origin is determined by its parametric equations $\{x_1 = \cos\tau, x_2 = \sin\tau, x_3 = 0\}$. Substitution of the parametrization functions into the unit circle equation $x_1^2 + x_2^2 = 1$ makes its identity.

Introduce a length parameter $s \in [0, |L|]$ of L which is called the *natural parameter*. Here, $|L|$ denotes the length of the support of the curve. It can be simply presented as a filament of length $|L|$ fitted into the support of $|L|$ as shown in Fig.7.11).

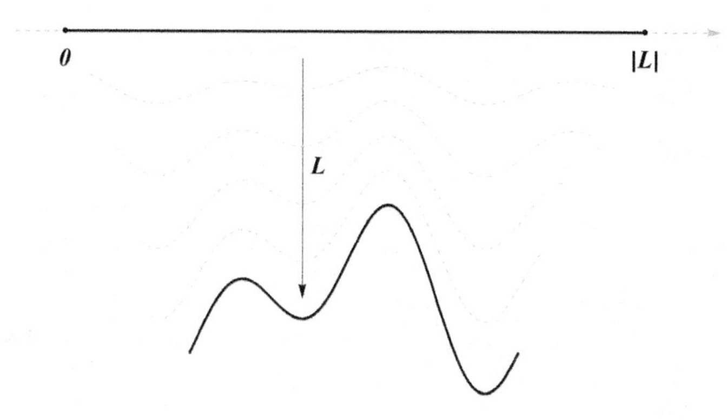

FIGURE 7.11: Illustration of the natural parametrization.

Let a unit vector $\mathbf{s}$ be tangent at each point to a smooth curve L. Then, the directional derivative $\frac{\partial u}{\partial \mathbf{s}}$ coincides with the parametric derivative $\frac{\partial u}{\partial s}$ on the natural parameter of L.

A smooth curve L is oriented if one of two possible natural parametrizations is fixed, i.e., a direction is assigned to L. Fix a point M on L. A unit vector $\mathbf{n} = (n_1, n_2, n_3)$ perpendicular to the tangent vector $\mathbf{s}$ is called the *unit normal vector* to L at a point M. Two unit normal vectors can be constructed at each point of L. For definiteness, the *exterior unit normal vector* to L is chosen in such a way that the pair $(\mathbf{s}, \mathbf{n})$ forms a local right basis coincided to the standard basis of the two-dimensional (2D) Cartesian system, i.e., $\mathbf{s} = (1, 0)$ and $\mathbf{n} = (0, 1)$ in local coordinates with the origin at the point M.

Let c denote a constant. Consider the equation $u(x_1, x_2) = c$ which determines a curve on the plane x_1, x_2 called the *level curve* (in general it can be the empty set or a point). If c changes over C the set $\{u(x_1, x_2) = c,\ c \in C\}$ is called a family of curves. An example of the family is shown in Fig.7.12.

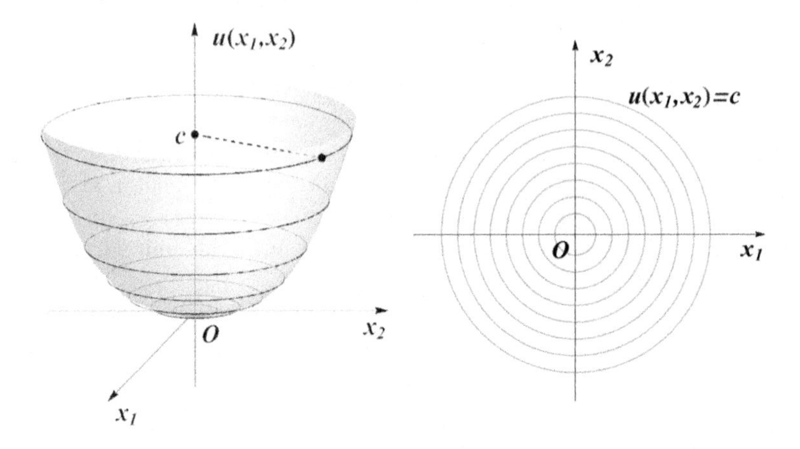

FIGURE 7.12: The level set of the function $u(x_1, x_2) = x_1^2 + x_2^2$ are circles $x_1^2 + x_2^2 = c$.

The scalar product (7.53) can be written in the form

$$\frac{\partial u}{\partial \mathbf{w}} = |\nabla u||\mathbf{w}| \cos \theta, \tag{7.55}$$

where θ denotes the angle between the vectors ∇u and $\mathbf{w}$. Consider a smooth level curve L and a point $\mathbf{x} \in L$ (Fig.7.13). Let $\mathbf{s}$ be the unit tangent vector and $\mathbf{n}$ be the unit normal vector to L at the point $\mathbf{x}$. The function $u = u(x_1, x_2)$ is constant on L, hence

$$\frac{\partial u}{\partial \mathbf{s}} = 0. \tag{7.56}$$

Equation (7.55) implies that the vectors $\mathbf{s}$ and ∇u are perpendicular. Therefore, the vector ∇u is parallel to $\mathbf{n}$ and

$$\frac{\partial u}{\partial \mathbf{n}} = \pm|\nabla u|. \tag{7.57}$$

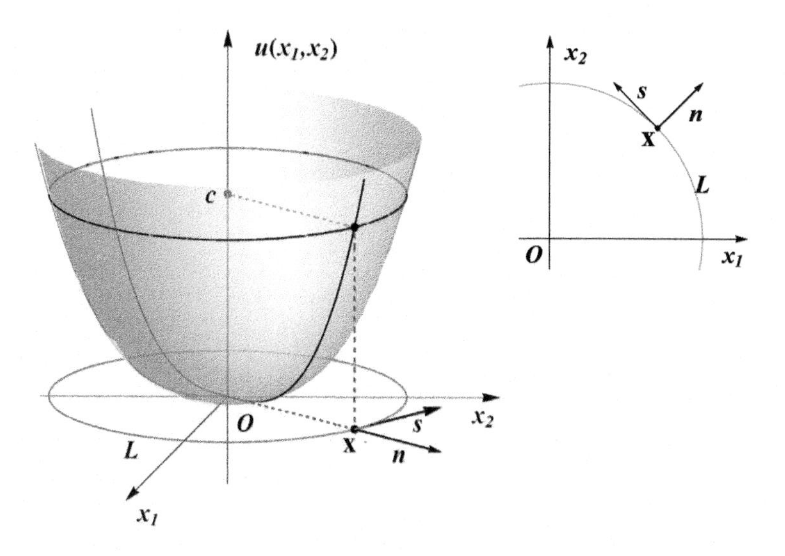

FIGURE 7.13: The unit tangent and normal vectors, **s** and **n**, respectively, to L at the point **x**. The direction of greatest increase of $u = u(x_1, x_2)$ at **x** coincides with ∇u. In the considered case, it is **n**; the direction of greatest decrease is $(-\mathbf{n})$. The function does not change in the direction $\pm\mathbf{s}$.

The sign plus (minus) is taken in (7.57) when the function $u = u(x_1, x_2)$ increases (decreases) in the direction **n**. The direction of greatest change of $u = u(x_1, x_2)$ coincides with the gradient. *The gradient descent method* is based on this observation. It is a numerical method to determine a local extremal point of a function. In order to determine a minimum, we arbitrarily fix a point **x** and descend step by step in the direction $-|\nabla u|$ to the local minimum.

It is worth noting that the families of level lines and of gradients generate two orthogonal nets on the plane (see Fig.7.14).

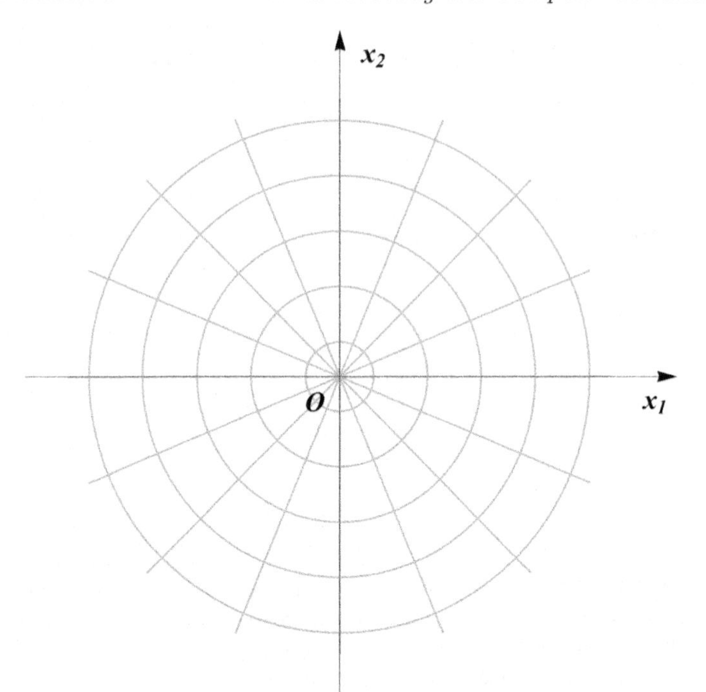

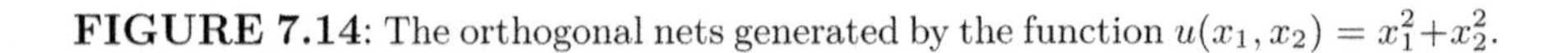

FIGURE 7.14: The orthogonal nets generated by the function $u(x_1, x_2) = x_1^2 + x_2^2$.

Similar results hold in $\mathbb{R}^3$. *Level surfaces* $u(x_1, x_2, x_3) = c$ are used instead of level lines.

The gradient was introduced with formula (7.51) in the Cartesian coordinates. The gradient, as a physical value having the tensor character, can be defined independently on coordinates. Such an alternative definition is given by the surface integral

$$\nabla u = \lim_{|D| \to 0} \frac{1}{|D|} \int_{\partial D} u \, \mathbf{n} \, dS, \tag{7.58}$$

where a domain D is bounded by the surface ∂D and $|D|$ denotes the volume of D.

7.6.2 Divergence

Let $\mathbf{a} = (a_1, a_2, a_3)$ be a continuously differentiable vector field. The *divergence* of $\mathbf{a}$ is defined as

$$\text{div } \mathbf{a} = \nabla \cdot \mathbf{a}. \tag{7.59}$$

In the Cartesian coordinates, we have

$$\text{div } \mathbf{a} = \nabla \cdot \mathbf{a} = \frac{\partial a_1}{\partial x_1} + \frac{\partial a_2}{\partial x_2} + \frac{\partial a_3}{\partial x_3}. \tag{7.60}$$

Analogously to (7.58) the divergence can be introduced independently on coordinates in terms of the surface integral

$$\nabla \cdot \mathbf{a} = \lim_{|D| \to 0} \frac{1}{|D|} \int_{\partial D} \mathbf{a} \cdot \mathbf{n} \, dS. \tag{7.61}$$

Consider the following equation near a point $\mathbf{x}$

$$\nabla \cdot \mathbf{a}(\mathbf{x}) = 0. \tag{7.62}$$

This equation implies that the vector field $\mathbf{a}$ does not possess sources and sinks near $\mathbf{x}$, i.e., $\mathbf{a}$ is similar to a laminar field. In the opposite case, the vector field $\mathbf{a}$ has a singularity at $\mathbf{x}$ (see Fig.7.15).

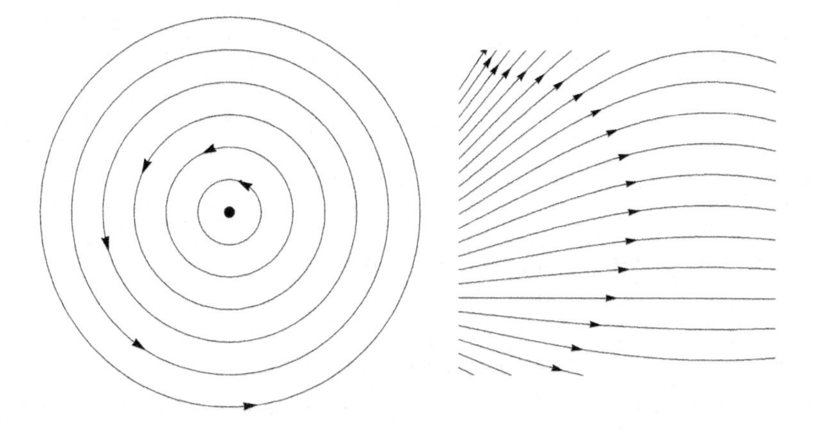

FIGURE 7.15: Singular and laminar fields.

For the detailed discussion of equation (7.62), we consider a heat flux $\mathbf{q} = (q_1, q_2, q_3)$ near $\mathbf{x}$. Let D_1 denote a 2D domain of the unit area (for instance, the unit square) which is perpendicular to the x_1-axis and contains the point $\mathbf{x}$. The component q_1 is the total heat passing through D_1 per time unit in the x_1-direction. The value $\frac{\partial q_1}{\partial x_1}$ expresses the spatial change of the heat in the x_1-direction near $\mathbf{x}$. In the 1D case, the heat flux has the form $\mathbf{q} = (q_1(x_1), 0, 0)$ and equation (7.62) becomes

$$\frac{dq_1}{dx_1} = 0. \tag{7.63}$$

Equation (7.63) demonstrates the Law of Conservation of Energy. The heat flux does not change near the point $\mathbf{x}$. In other words, the input heat is equal to the output heat at $\mathbf{x}$.

The same arguments hold in the 3D case. The difference is that the heat flux must be balanced in all the directions. The values $\frac{\partial q_2}{\partial x_2}$ and $\frac{\partial q_3}{\partial x_3}$ express the spacial change in the x_2- and x_3- directions. The total spatial change of the heat flux at $\mathbf{x}$ is equal to

$$\frac{\partial q_1}{\partial x_1} + \frac{\partial q_2}{\partial x_2} + \frac{\partial q_3}{\partial x_3}.$$

If the heat does not disappear and does not arise at the point $\mathbf{x}$, the energy conservation law implies that the latter value must vanish. Conversely, if the divergence vanishes at $\mathbf{x}$, then the heat balance becomes zero. That is, the heat does not disappear and does not arise at $\mathbf{x}$.

Example 7.5. Consider the heat flux

$$\mathbf{q}(\mathbf{x}) = q_0 \frac{\mathbf{x}}{r^3} = q_0 \frac{1}{r^3}(x_1, x_2, x_3), \tag{7.64}$$

where q_0 is a constant and $r = |\mathbf{x}|$. Using the relation $\frac{\partial r}{\partial x_i} = \frac{x_i}{r}$ $(i = 1, 2, 3)$ we calculate the derivatives

$$\frac{\partial q_i}{\partial x_i} = q_0 \frac{\partial}{\partial x_i}\left(\frac{x_i}{r^3}\right) = q_0 \left(\frac{1}{r^3} - \frac{3x_i^2}{r^5}\right). \tag{7.65}$$

Therefore, everywhere except at the origin we have

$$\nabla \cdot \mathbf{q} = 0. \tag{7.66}$$

The vector field (7.64) is singular at $\mathbf{x} = \mathbf{0}$. Such a singularity will be discussed in Example 7.7 on page 161.

MATLAB Example Box 7.1

Let us visualize the vector field with sources and sinks. One can consider (7.64) as the field of a point source ($q_0 > 0$) or the field of a point sink ($q_0 < 0$). The constant q_0 is proportional to the *strength* of the source. As an example, we construct the following field incorporating n points $\mathbf{x}_i$ described by series q_i $(i = 1, 2, 3, \ldots, n)$:

$$\mathbf{q}(\mathbf{x}) = \sum_{i=1}^{n} q_i \frac{\mathbf{x} - \mathbf{x}_i}{|\mathbf{x} - \mathbf{x}_i|^3}.$$

As an example, consider a vector field incorporating two sinks at points $(-2, -0.7)$, $(0.5, 2)$, as well as a single source at $(2, -2)$.

```
function script16()

  % define sources and sinks
  xk = [-2     2  0.5];
  yk = [-0.7 -2   2  ];
  qk = [-1    1  -1  ];

  % generate the mesh of points
  x = linspace(-3, 3, 31);
  y = linspace(-3, 3, 31);
```

```
[X, Y] = meshgrid(x, y);

% compute values of vectors at (X, Y)
U = zeros(length(X));
V = zeros(length(X));
for k = 1:length(qk)
    d = sqrt( ((X-xk(k)).^2) + ((Y-yk(k)).^2) );
    U = U + qk(k) * (X - xk(k)) ./ d.^3 ;
    V = V + qk(k) * (Y - yk(k)) ./ d.^3 ;
end

% normalize vectors
d = sqrt( U.^2 + V.^2 );
U = U ./ d;
V = V ./ d;

% plot the vector field
hold on
quiver(X, Y, U, V)
plot(xk, yk, 'or')
hold off
grid on
axis equal
axis([-3 3 -3 3])
```

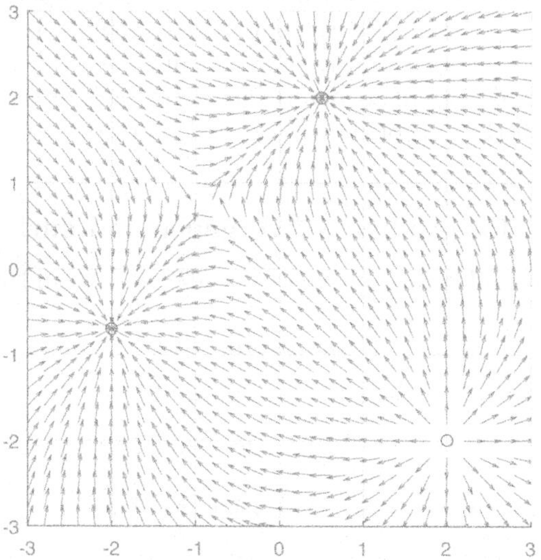

In order to visualize the field, the `quiver` operator was used. It plots vectors with components (`U`, `V`) at the points (`X`, `Y`). Here, the vectors are normalized. In order to add new plots to the current figure, instead of replacing the existing plots, activate the `hold` option.

7.6.3 Curl

The *curl* of the continuously differentiable vector field $\mathbf{a}$ is introduced as the vector

$$\text{rot } \mathbf{a} = \nabla \times \mathbf{a}. \tag{7.67}$$

The notation curl is also used. The curl operator in *Mathematica* is denoted by `Curl`. In the Cartesian coordinates,

$$\text{rot } \mathbf{a} = \text{curl } \mathbf{a} = \nabla \times \mathbf{a} = \begin{vmatrix} \mathbf{i}_1 & \mathbf{i}_2 & \mathbf{i}_3 \\ \frac{\partial}{\partial x_1} & \frac{\partial}{\partial x_2} & \frac{\partial}{\partial x_3} \\ a_1 & a_2 & a_3 \end{vmatrix}. \tag{7.68}$$

Calculation of the determinant yields

$$\nabla \times \mathbf{a} = \left(\frac{\partial a_3}{\partial x_2} - \frac{\partial a_2}{\partial x_3}, \frac{\partial a_1}{\partial x_3} - \frac{\partial a_3}{\partial x_1}, \frac{\partial a_2}{\partial x_1} - \frac{\partial a_1}{\partial x_2} \right). \tag{7.69}$$

The curl can be introduced independently on coordinates similar to (7.58) and (7.61):

$$\nabla \times \mathbf{a} = \lim_{|D| \to 0} \frac{1}{|D|} \int_{\partial D} \mathbf{n} \times \mathbf{a} \, dS. \tag{7.70}$$

Example 7.6. The rotation of a body about the origin is described in Sec.7.3. The linear velocity at a point $\mathbf{x}$ is calculated with formula $\mathbf{v} = \boldsymbol{\omega} \times \mathbf{x}$, where ω denotes the angle velocity. Calculate the curl of the vector field as $\nabla \times \mathbf{v} = \nabla \times (\boldsymbol{\omega} \times \mathbf{x}) = 2\boldsymbol{\omega}$. This formula explains the vector form of the angle velocity which coincides with the curl of the vector field up to the multiplier 2.

7.6.4 Formulae for gradient, divergence and curl

The following formulae take place for a scalar field φ and vector fields $\mathbf{a}$, $\mathbf{b}$ [13]

$$\nabla \times \nabla\varphi = 0, \tag{7.71}$$

$$\nabla \cdot (\mathbf{a} \times \mathbf{b}) = \mathbf{b}(\nabla \times \mathbf{a}) - \mathbf{a}(\nabla \times \mathbf{b}), \tag{7.72}$$

$$\nabla \times (\varphi\mathbf{a}) = \varphi(\nabla \times \mathbf{a}) - \mathbf{a}(\nabla\varphi), \tag{7.73}$$

$$\frac{1}{2}\nabla|\mathbf{a}|^2 = \mathbf{a} \times (\nabla \times \mathbf{a}) + (\mathbf{a} \cdot \nabla)\mathbf{a}, \tag{7.74}$$

$$\nabla \times (\mathbf{a} \times \mathbf{b}) = (\mathbf{a} \cdot \nabla)\mathbf{b} - (\mathbf{b} \cdot \nabla)\mathbf{a} + \mathbf{a}\nabla \cdot \mathbf{b} - \mathbf{b}\nabla \cdot \mathbf{a}. \tag{7.75}$$

Let these fields have the form $\varphi = \varphi[f(\mathbf{x})]$, $\mathbf{a} = \mathbf{a}[f(\mathbf{x})]$, where f is a function of one variable. Then, we have

$$\nabla\varphi[f(\mathbf{x})] = \frac{d\varphi}{df}[f(\mathbf{x})] \, \nabla f(\mathbf{x}), \tag{7.76}$$

$$\nabla \cdot \mathbf{a}[f(\mathbf{x})] = \frac{d\mathbf{a}}{df}[f(\mathbf{x})] \cdot \nabla f(\mathbf{x}), \tag{7.77}$$

$$\nabla \times \mathbf{a}[f(\mathbf{x})] = \nabla f(\mathbf{x}) \times \frac{d\mathbf{a}}{df}[f(\mathbf{x})], \tag{7.78}$$

$$\nabla^2 \mathbf{a} = \nabla(\nabla \cdot \mathbf{a}) - \nabla \times (\nabla \times \mathbf{a}). \tag{7.79}$$

For instance, (7.76) yields

$$\nabla|\mathbf{x}| = \frac{\mathbf{x}}{|\mathbf{x}|}, \quad \nabla\frac{1}{|\mathbf{x}|} = -\frac{\mathbf{x}}{|\mathbf{x}|^3}. \tag{7.80}$$

Formulae for the basic differential operators in other coordinates are useful in applications. Such forulae are implemented in *Mathematica*. For the cylindrical coordinates (see Sec.7.1.2), we have

```
In[1]:= Needs["VectorAnalysis`"]

In[2]:= SetCoordinates[Cylindrical]

Out[2]= Cylindrical[Rr, Ttheta, Zz]

In[3]:= CoordinateRanges[]

Out[3]= {0 ≤ Rr < ∞, -π < Ttheta ≤ π, -∞ < Zz < ∞}

In[4]:= JacobianDeterminant[]

Out[4]= Rr

In[5]:= Grad[f[Rr, Ttheta, Zz]] /. {Rr → r, Ttheta → θ, Zz → z}

Out[5]= {f^(1,0,0)[r, θ, z], f^(0,1,0)[r, θ, z]/r, f^(0,0,1)[r, θ, z]}

In[6]:= a[Rr_, Ttheta_, Zz_] = {a1[Rr, Ttheta, Zz], a2[Rr, Ttheta, Zz],
          a3[Rr, Ttheta, Zz]};

In[7]:= Div[a[Rr, Ttheta, Zz]] /. {Rr → r, Ttheta → θ, Zz → z} //
          Simplify

Out[7]= 1/r (a1[r, θ, z] + r a3^(0,0,1)[r, θ, z] +
          a2^(0,1,0)[r, θ, z] + r a1^(1,0,0)[r, θ, z])
```

```
In[8]:= Curl[a[Rr, Ttheta, Zz]] /. {Rr → r, Ttheta → θ, Zz → z} //
        Simplify
```

$$\text{Out[8]=} \left\{-a2^{(0,0,1)}[r,\theta,z]+\frac{a3^{(0,1,0)}[r,\theta,z]}{r},\ a1^{(0,0,1)}[r,\theta,z]-a3^{(1,0,0)}[r,\theta,z],\ \frac{a2[r,\theta,z]-a1^{(0,1,0)}[r,\theta,z]+r\,a2^{(1,0,0)}[r,\theta,z]}{r}\right\}$$

```
In[9]:= Laplacian[f[Rr, Ttheta, Zz]] /. {Rr → r, Ttheta → θ, Zz → z} //
        Simplify
```

$$\text{Out[9]=} f^{(0,0,2)}[r,\theta,z]+\frac{f^{(0,2,0)}[r,\theta,z]}{r^2}+\frac{f^{(1,0,0)}[r,\theta,z]}{r}+f^{(2,0,0)}[r,\theta,z]$$

The latter formula in the traditional form becomes

$$\nabla^2 f = \frac{1}{r}\frac{\partial}{\partial r}\left(r\frac{\partial f}{\partial r}\right) + \frac{1}{r^2}\frac{\partial^2 f}{\partial \theta^2} + \frac{\partial^2 f}{\partial z^2}. \tag{7.81}$$

For the spherical coordinates (see Sec.7.1.3), we have

```
In[1]:= Needs["VectorAnalysis`"]

In[2]:= SetCoordinates[Spherical]

Out[2]= Spherical[Rr, Ttheta, Pphi]

In[3]:= CoordinateRanges[]

Out[3]= {0 ≤ Rr < ∞, 0 ≤ Ttheta ≤ π, -π < Pphi ≤ π}

In[4]:= JacobianDeterminant[]

Out[4]= Rr^2 Sin[Ttheta]

In[5]:= Grad[f[Rr, Ttheta, Pphi]] /. {Rr → r, Ttheta → θ, Pphi → ϕ}
```

$$\text{Out[5]=} \left\{f^{(1,0,0)}[r,\theta,\phi],\ \frac{f^{(0,1,0)}[r,\theta,\phi]}{r},\ \frac{\text{Csc}[\theta]\,f^{(0,0,1)}[r,\theta,\phi]}{r}\right\}$$

```
In[6]:= a[Rr_, Ttheta_, Pphi_] = {a1[Rr, Ttheta, Pphi], a2[Rr, Ttheta, Pphi],
          a3[Rr, Ttheta, Pphi]};

In[7]:= Div[a[Rr, Ttheta, Pphi]] /. {Rr → r, Ttheta → θ, Pphi → ϕ} // Simplify
```

$$\text{Out[7]=} \frac{1}{r}\left(2\,a1[r,\theta,\phi]+a2[r,\theta,\phi]\,\text{Cot}[\theta]+\text{Csc}[\theta]\,a3^{(0,0,1)}[r,\theta,\phi]+a2^{(0,1,0)}[r,\theta,\phi]+r\,a1^{(1,0,0)}[r,\theta,\phi]\right)$$

```
In[8]:= Curl[a[Rr, Ttheta, Pphi]] /. {Rr → r, Ttheta → θ, Pphi → ϕ} // Simplif
```

Out[8]= $\left\{\frac{\text{a3}[r,\theta,\phi]\,\text{Cot}[\theta]-\text{Csc}[\theta]\,\text{a2}^{(0,0,1)}[r,\theta,\phi]+\text{a3}^{(0,1,0)}[r,\theta,\phi]}{r},\right.$
$-\frac{\text{a3}[r,\theta,\phi]-\text{Csc}[\theta]\,\text{a1}^{(0,0,1)}[r,\theta,\phi]+r\,\text{a3}^{(1,0,0)}[r,\theta,\phi]}{r},$
$\left.\frac{\text{a2}[r,\theta,\phi]-\text{a1}^{(0,1,0)}[r,\theta,\phi]+r\,\text{a2}^{(1,0,0)}[r,\theta,\phi]}{r}\right\}$

```
In[9]:= Laplacian[f[Rr, Ttheta, Pphi]] /. {Rr → r, Ttheta → θ, Pphi → ϕ} //
        Simplify
```

Out[9]= $\frac{1}{r^2}\left(\text{Csc}[\theta]^2\,f^{(0,0,2)}[r,\theta,\phi]+\text{Cot}[\theta]\,f^{(0,1,0)}[r,\theta,\phi]+ f^{(0,2,0)}[r,\theta,\phi]+2\,r\,f^{(1,0,0)}[r,\theta,\phi]+r^2\,f^{(2,0,0)}[r,\theta,\phi]\right)$

Similar computations can be performed in other coordinates implemented in *Mathematica* (see Help: `VectorAnalysis/tutorial/VectorAnalysis`).

7.7 Integral theorems

The standard course of vector calculus includes the following integral theorems.

Theorem 7.1 (Ostrogradsky-Gauss[6]). *Let continuously differentiable scalar functions $P(\mathbf{x})$, $Q(\mathbf{x})$ and $R(\mathbf{x})$ be given in the domain $D \cup \partial D \subset \mathbb{R}^3$ with the piecewise smooth boundary surface ∂D. Then,*

$$\begin{aligned}&\int_D \left(\frac{\partial P}{\partial x_1}+\frac{\partial Q}{\partial x_2}+\frac{\partial R}{\partial x_3}\right) d\mathbf{x}\\ &= \int_{\partial D} (P\cos(\mathbf{n},x_1)+Q\cos(\mathbf{n},x_2)+R\cos(\mathbf{n},x_3))\,dS,\end{aligned} \tag{7.82}$$

where $(\mathbf{n}, x_i)$ denotes the angle between the vector $\mathbf{n}$ normal to ∂D and the x_i-axis.

Equation (7.82) relates triple and surface integrals.

Theorem 7.2 (Stokes). *Let S be a smooth oriented surface in the space $\mathbb{R}^3$ with the closed regular and smooth boundary curve ∂S. Let scalar functions*

---

[6]Frequently, it is called Gauss's theorem. *Arnold's Principle* : If a notion bears a personal name, then this name is not the name of the discoverer. *Berry Principle*: Arnold's Principle is applicable to itself.

$P(\mathbf{x})$, $Q(\mathbf{x})$ and $R(\mathbf{x})$, and their derivatives $\frac{\partial P}{\partial x_2}(\mathbf{x})$, $\frac{\partial P}{\partial x_3}(\mathbf{x})$, $\frac{\partial Q}{\partial x_1}(\mathbf{x})$, $\frac{\partial Q}{\partial x_3}(\mathbf{x})$, $\frac{\partial R}{\partial x_1}(\mathbf{x})$, $\frac{\partial R}{\partial x_2}(\mathbf{x})$ are continuous on $S \cup \partial S$. Then,

$$\begin{aligned} &\int_S \left[\left(\frac{\partial R}{\partial x_2} - \frac{\partial Q}{\partial x_3}\right)\cos(\mathbf{n}, x_1) + \left(\frac{\partial P}{\partial x_3} - \frac{\partial R}{\partial x_1}\right)\cos(\mathbf{n}, x_2)\right. \\ &\left. + \left(\frac{\partial Q}{\partial x_1} - \frac{\partial P}{\partial x_1}\right)\cos(\mathbf{n}, x_3)\right] dS = \int_{\partial D} P dx_1 + Q dx_2 + R dx_3, \end{aligned} \tag{7.83}$$

where $\mathbf{n}$ denotes the outward normal vector to the surface S.

Equation (7.83) relates surface and curvilinear integrals.

Theorem 7.3 (Green). *Let continuously differentiable scalar functions $P(\mathbf{x})$, $Q(\mathbf{x})$ be given in the 2D domain $D \cup \partial D \subset \mathbb{R}^2$ with the piecewise smooth simple boundary curve ∂D. Then,*

$$\int_{\partial D} P dx_1 + Q dx_2 = \int_D \left(\frac{\partial Q}{\partial x_1} - \frac{\partial P}{\partial x_2}\right) d\mathbf{x}. \tag{7.84}$$

Equation (7.84) relates curvilinear and double integrals.

It is possible to rewrite formulae (7.82) and (7.83) in terms of the vector analysis. Introduce the vector function $\mathbf{a} = (P, Q, R)$. Then, the integrand from the left-hand side of (7.82) is the divergence of the vector field $\mathbf{a}$, hence it is equal to $\nabla \cdot \mathbf{a}$. The integrand from the left-hand side of (7.82) can be treated as the scalar product of the vectors $\mathbf{a}$ and $\mathbf{n}$. Therefore, Ostrogradsky-Gauss's formula becomes

$$\int_D \nabla \cdot \mathbf{a}\, d\mathbf{x} = \int_{\partial D} \mathbf{a} \cdot \mathbf{n}\, ds. \tag{7.85}$$

Along similar lines Stokes's formula (7.83) becomes

$$\int_\Gamma \mathbf{a} \cdot d\mathbf{s} = \int_S (\nabla \times \mathbf{a}) \cdot d\mathbf{S}, \tag{7.86}$$

where $d\mathbf{S} = \mathbf{n}\, ds$.

Put $\mathbf{a} = (\varphi, 0, 0)$, $\mathbf{a} = (0, \varphi, 0)$, $\mathbf{a} = (0, 0, \varphi)$ one by one in (7.85). Write the obtained three scalar formulae as a vector one

$$\int_D \nabla\varphi\, d\mathbf{x} = \int_{\partial D} \varphi\mathbf{n}\, ds. \tag{7.87}$$

Let u and v be functions twice continuously differentiable in D and continuously differentiable in its closure. Application of (7.85) to $\mathbf{a} = v\nabla u$ yields Green's formula

$$\int_D (\nabla v \cdot \nabla u + v\nabla^2 u) d\mathbf{x} = \int_{\partial D} v \frac{\partial u}{\partial n}\, ds. \tag{7.88}$$

Example 7.7. Let the heat flux from Example 7.5 be presented as the vector field $\mathbf{q} = \frac{\mathbf{x}}{|\mathbf{x}|^3}$ from (7.64). Everywhere $\nabla \cdot \mathbf{q} = 0$ except the point $\mathbf{x} = \mathbf{0}$, where the vectro field $\mathbf{q}$ is singular. In order to investigate this singularity we surround the point by a sphere S_0 of radius r_0 with center at $\mathbf{x} = \mathbf{0}$ and calculate the total heat flux across the sphere S_0

$$\int_{S_0} \mathbf{q} \cdot \mathbf{n}\, ds = q_0 \int_{S_0} \frac{\mathbf{x} \cdot \mathbf{x}}{r_0^4}\, ds = 4\pi q_0, \tag{7.89}$$

where the relation $\mathbf{x} \cdot \mathbf{x} = r_0^2$ is used. The area of S_0 holds $\int_{S_0} ds$ and it is equal to $4\pi r_0^2$. One can see that the result (7.89) does not depend on the radius r_0 which can be taken infinitesimally small.

Let us formally apply Ostrogradsky-Gauss's formula to the singular vector function $\mathbf{q}$

$$\int_{D_0} \nabla \cdot \mathbf{q}\, d\mathbf{x} = 4\pi q_0 \tag{7.90}$$

for the ball D_0 centered at the origin with a radius r_0. A similar result was obtained in Sec.6.5 by use of the δ-function

$$4\pi q_0 \int_{D_0} \delta(\mathbf{x})\, d\mathbf{x} = 4\pi q_0. \tag{7.91}$$

Hence, the value $\nabla \cdot \mathbf{q}$ can be calculated for any $\mathbf{x}$ including the singular point $\mathbf{x} = \mathbf{0}$

$$\nabla \cdot \left(\frac{\mathbf{x}}{|\mathbf{x}|^3} \right) = 4\pi \delta(\mathbf{x}). \tag{7.92}$$

Moreover, the second formula (7.80) implies that

$$\frac{\mathbf{x}}{|\mathbf{x}|^3} = -\nabla \frac{1}{|\mathbf{x}|}. \tag{7.93}$$

Substitution of (7.93) into (7.92) yields the following formula for Laplace's operator applied to $|\mathbf{x}|^{-1}$

$$\nabla^2 \frac{1}{|\mathbf{x}|} = -4\pi \delta(\mathbf{x}). \tag{7.94}$$

Therefore, the function $u(\mathbf{x}) = |\mathbf{x}|^{-1}$ satisfies the Laplace equation everywhere except zero where an infinite jump occurs.

Further reading. We recommend the books [13, 16, 52].

Exercises

1. Prove the identities (7.28)-(7.32) and (7.71)-(7.79).

2. Prove that

   $\nabla \log |\mathbf{x}| = \frac{\mathbf{x}}{|\mathbf{x}|^2}$;

   $\nabla \cdot \mathbf{x} = 3$ in $\mathbb{R}^3$;

   $\nabla \times \mathbf{x} = 0$.

3. A vector field $\mathbf{a}(\mathbf{x})$ is called potential, if it is the gradient of a scalar field $\varphi(\mathbf{x})$, i.e., $\mathbf{a}(\mathbf{x}) = \nabla\varphi(\mathbf{x})$ in a simply connected domain $D \subset \mathbb{R}^3$. Prove that if $\mathbf{a}(\mathbf{x})$ is potential, then $\nabla \times \mathbf{a} = \mathbf{0}$.

   Hint: Calculate $\nabla \times \nabla\varphi(\mathbf{x})$ with formula (7.69).

   It can be proved that $\mathbf{a}(\mathbf{x})$ is potential if and only if $\nabla \times \mathbf{a} = \mathbf{0}$ (see for details [13]).

   Give an example that it is not true for multiply connected domains, for instance for an annulus.

4. A vector field $\mathbf{a}(\mathbf{x})$ is called solenoidal, if it is the curl of another vector field $\mathbf{b}(\mathbf{x})$, i.e., $\mathbf{a}(\mathbf{x}) = \nabla \times \mathbf{b}(\mathbf{x})$. Prove that if $\mathbf{a}(\mathbf{x})$ is solenoidal, then $\nabla \cdot \mathbf{a} = 0$.

   It can be proved that $\mathbf{a}(\mathbf{x})$ is solenoidal if and only if $\nabla \cdot \mathbf{a} = \mathbf{0}$ (see [13]).

Chapter 8

Heat equations

8.1 Heat conduction equations 163
8.2 Initial and boundary value problems 168
8.3 Green's function for the 1D heat equation 169
8.4 Fourier series 173
8.5 Separation of variables 177
8.6 Discrete approximations of PDE 180
8.6.1 Finite-difference method 180
8.6.2 1D finite element method 182
8.6.3 Finite element method in $\mathbb{R}^2$ 186
8.7 Universality in Mathematical Modeling. Table 188
Exercises 189

In the present chapter, mathematical models for the heat conduction are developed. Particular 1D problems were discussed in Chapter 6. The heat conduction is taken for definiteness. The same equation, here called the heat equation, describes diffusion, electrostatics etc. (see Batchelor's Table in Sec.8.7). It is not surprising that this equation is also called the diffusion equation. This is the standard situation in Mathematical Modeling when processes from different topics of science are described by the same equation. Roughly speaking the automatic replacements: "heat conduction" → "diffusion", "heat flux" → "diffusion flux", "thermal conductivity" → "diffusivity" etc. transforms a course on the heat conduction to a course on the diffusion with the fitting 90%. At least a short comparative review of the both topics could be useful to argue with the above assertion.

8.1 Heat conduction equations

The mathematical model for the heat conduction relates two fundamental values: the temperature distribution represented by a scalar function $u = u(\mathbf{x}, t)$ and the heat flux represented by a vector-function $\mathbf{q}(\mathbf{x}) = (q_1(\mathbf{x}, t), q_2(\mathbf{x}, t), q_3(\mathbf{x}, t))$. These functions are determined in a domain $D \subset \mathbb{R}^3$ bounded by a piecewise smooth surface ∂D, i.e., $\mathbf{x} = (x_1, x_2, x_3) \in D \cup \partial D$.

1D stationary heat conduction was discussed in Chapter 6. In the present section, first we consider non stationary processes in an isotropic conducting medium (the term *isotropy* is precisely explained below after equation (8.3)). The temperature distribution $u = u(\mathbf{x}, t)$ and the first component of the heat flux $q_1(\mathbf{x}, t)$ satisfy the Fourier law in the x_1-direction, c.f. (6.3),

$$q_1 = -\lambda \frac{\partial u}{\partial x_1}, \tag{8.1}$$

where the constant λ is called the conductivity and depends on the medium. Its measure is $[\lambda] = \frac{W}{mK}$ (Watt per Meter-Kelvin) in the SI units.

The Fourier law holds in two other directions[1]

$$q_2 = -\lambda \frac{\partial u}{\partial x_2}, \quad q_3 = -\lambda \frac{\partial u}{\partial x_3}. \tag{8.2}$$

Three scalar equations (8.1)-(8.2) can be written in the vector form

$$\mathbf{q} = -\lambda \nabla u. \tag{8.3}$$

This is the 3D Fourier law for isotropic media. More precisely, the medium is called *isotropic* if the heat flux and the temperature gradient are related by (8.3) with a proportionality coefficient $(-\lambda)$. It is worth noting that λ can be a function of $\mathbf{x}$, i.e., the conduction properties of the medium can change in space.

The Fourier law for anisotropic media has the form

$$\mathbf{q} = -\Lambda \nabla u, \tag{8.4}$$

where Λ is a tensor of second order. In particular, if $\Lambda = \lambda \mathbf{I}$ where $\mathbf{I}$ is the identity matrix, then (8.4) becomes (8.3), i.e., the considered medium is isotropic. The general physical properties of the heat conduction imply that the tensor Λ is always represented by means of symmetric positively determined matrices in the Cartesian coordinates. The anisotropy effect becomes clear in such coordinates where the tensor Λ is represented by a diagonal matrix[2]

$$\Lambda = \begin{pmatrix} \lambda_1 & 0 & 0 \\ 0 & \lambda_2 & 0 \\ 0 & 0 & \lambda_3 \end{pmatrix} \tag{8.5}$$

The vector equation (8.4) can be written by coordinates

$$q_1 = -\lambda_1 \frac{\partial u}{\partial x_1}, \quad q_2 = -\lambda_2 \frac{\partial u}{\partial x_2}, \quad q_3 = -\lambda_3 \frac{\partial u}{\partial x_3}. \tag{8.6}$$

[1] It is valid in any direction.

[2] It follows from linear algebra that any symmetric positively determined matrix can be reduced to a diagonal one. One can find applications of matrix analysis to tensors of second and higher orders in [13].

These equations demonstrate that the heat conduction is different in different directions than what characterizes *anisotropy* of medium.

The second fundamental law of heat transfer concerns *heat capacity* which is defined as the energy (heat) necessary to heat (to cool) up to 1^0 a unit mass of material. The heat capacity is measured in $[c] = \frac{J}{kg\,K}$ (Joule per kilogram-Kelvin) in SI units. Then, the heat value Q necessary to heat a mass m up to Δu degree is calculated by formula

$$Q = cm\Delta u = c\rho V \Delta u, \tag{8.7}$$

where ρ denotes the density of material and V the volume of the considered sample.

We are now ready to deduce the heat equation using the principle of transition *continuous* $\leftrightarrow$ *discrete*. First, we consider the balance of heat of a rectangular cuboid in the x_1–direction with faces parallel to the coordinate axes. Let the cuboid have the width $\Delta x = b - a$ in the x_1-direction and the area S of the faces perpendicular to the axis x_1 (see Fig.8.1). Let the heat value Q_1 pass through the face $x_1 = a$ during the time Δt in the x_1-direction and the heat value Q_2 through the face $x_1 = b$. The balance of heat passing through the medium between $x_1 = a$ and $x_1 = b$ is calculated by formula

$$Q_1 - Q_2 = -S\Delta t \Delta q, \tag{8.8}$$

where Δq denotes the increment of the first coordinate of the heat flux between $x_1 = a$ and $x_1 = b$. The sign minus in (8.8) corresponds to the direction of the heat flux.

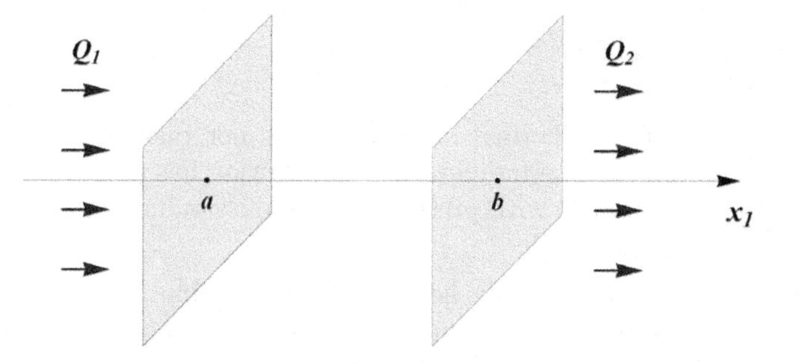

FIGURE 8.1: Heat flux in the x_1-direction.

From the other side, the same heat $Q_1 - Q_2$ is dissipated by the medium in accordance with formula (8.7)

$$Q_1 - Q_2 = c\rho S \Delta x \Delta u, \tag{8.9}$$

where $V = S\Delta x$ is the volume of a cuboid. The conservation energy law implies that

$$-\Delta t \Delta q = c\rho \Delta x \Delta u, \text{ hence } -\frac{\Delta q}{\Delta x} = c\rho \frac{\Delta u}{\Delta t}. \tag{8.10}$$

Following the principle of transition *continuous* $\leftrightarrow$ *discrete* we calculate the limit in (8.10) as $\Delta x \to 0$ and $\Delta t \to 0$ and obtain

$$-\frac{\partial q}{\partial x_1} = c\rho\frac{\partial u}{\partial t}. \tag{8.11}$$

Equation (8.11) expresses the heat balance for the flux $\mathbf{q} = (q_1, q_2, q_3)$ when $q_1 = q$, $q_2 = q_3 = 0$, i.e., the heat conduction occurs only in the x_1-direction. The general 3D heat transfer is based on the heat balance for all three axes. The extension of equation (8.11) to 3D has the form

$$-\left(\frac{\partial q_1}{\partial x_1} + \frac{\partial q_2}{\partial x_1} + \frac{\partial q_3}{\partial x_1}\right) = c\rho\frac{\partial u}{\partial t} \quad \Longleftrightarrow \quad -\nabla \cdot \mathbf{q} = c\rho\frac{\partial u}{\partial t}. \tag{8.12}$$

Substitution of (8.3) into (8.12) yields the *heat equation*

$$\frac{\partial u}{\partial t} = a^2\nabla^2 u, \tag{8.13}$$

where the constant $a = \sqrt{\frac{\lambda}{c\rho}}$ is called the *thermal diffusivity.*

The heat equation models the temperature distribution $u(\mathbf{x}, t)$ in space and in time. Consider the *stationary heat conduction* when the temperature distribution does not depend on time, hence $u = u(\mathbf{x})$. This means that the exchange of thermal energy takes place in time but it does not change in time. For instance, Fig.6.6 displays the stationary temperature distribution in the wall. The heat equation (8.13) in the stationary case becomes an equation called the *Laplace equation*

$$\nabla^2 u = 0. \tag{8.14}$$

One can see that the thermal diffusivity does not take part in the stationary heat conduction. The temperature distribution has to satisfy the Laplace equation in a domain D with prescribed boundary values on ∂D discussed in the next subsection.

Example 8.1. Consider a long hollow cylindrical pipe filled with hot water. Let the temperature of water and the environment be given as constants T_1 and T_2, respectively. Let the thermal conductivity of pipe hold λ. Find the temperature distribution in the pipe.

First, following the principle of the simplest model stated on page 12 we try to simplify the problem. It is clear that the considered problem is stationary since the exterior values T_1 and T_2 do not change in time. Therefore, we consider the Laplace equation (8.14), not the general equation (8.13). This removes the question "when" discussed at the beginning of the book in Sec.1.1.

The next question "where" concerns coordinate systems. One can introduce the Cartesian coordinates in such a way that the pipe axis will coincide with x_3.

However, the principle of an apple stated on page 100 can help us to introduce more convenient coordinates. The function $u(\mathbf{x})$ depends on two variables $\mathbf{x} = (x_1, x_2)$ in the Cartesian coordinates. One can associate the pipe shape with a cylinder and the cylindrical coordinates described in Sec.7.1.2. Another association can be related to the observation that u depends only on the distance $r = \sqrt{x_1^2 + x_2^2}$ that also leads to the cylindrical coordinates. Ultimately, we have to write the Laplace equation (8.14) in the cylindrical coordinates. It was already done in the previous section. It follows from (7.81) that the function $u = u(r)$ satisfies equation

$$\frac{1}{r}\frac{d}{dr}\left(r\frac{du}{dr}\right) = 0, \quad r_1 < r < r_2, \tag{8.15}$$

where r_1 and r_2 denote the interior and exterior radii of the pipe, respectively. The general solution of the ODE has the form

$$u(r) = C_1 \log r + C_2, \tag{8.16}$$

where C_1 and C_2 are arbitrary constants. Substitution of the boundary conditions $u(r_1) = T_1$ and $u(r_2) = T_2$ yields

$$\frac{(T_1 - T_2)\log r + T_2 \log r_1 - T_1 \log r_2}{\log \frac{r_1}{r_2}}. \tag{8.17}$$

The heat equation (8.13) has to be modified if singularities (sources and sinks) are located in the medium. Introduce the function $w = w(\mathbf{x}, t)$ which expresses the distribution of going in and out heat per unit volume and per unit time near the point $\mathbf{x}$. Then, we arrive at the *heat equation with sources and sinks*

$$\frac{\partial u}{\partial t} = a^2 \nabla^2 u + \frac{w}{c\rho}. \tag{8.18}$$

Remark 8.1. The heat equations (8.14) and (8.18) model the instant heat transfer (with infinite velocity). Here, the velocity of heat transfer and the velocity of front expansion $u(\mathbf{x}, t) = const$ have different meanings. Finite velocity of the heat transfer implies the relaxation phenomenon [42].

The following table contains the thermal property constants of some materials in SI units [13, p. 269]:

material	λ	c	temperature (K)
oxygen (O_2)	0.01833	911	200
	0.02657	920	300
water (H_2O)	0.58	4.183	300
sodium (Na)	86.2	13.8	366
	72.8	13.0	644
glass	1.05	0.84	300
wood (along the axis of tree)	0.126	–	–
wood (across the axis of tree)	0.038	–	–

8.2 Initial and boundary value problems

Solutions of equations (8.14) and (8.18) in a domain D describe eventual temperature distributions. The general solution can be presented as a finite linear combination with undetermined constants as in ODE only in exceptional cases reduced to 1D. This feature essentially complicates solutions to PDE. Moreover, each distribution depends on the initial and boundary values. The temperature distribution at the beginning time $t = 0$ is called the *initial temperature.* It can be given as an arbitrary function $f(\mathbf{x})$ (for instance continuous in $D \cup \partial D$). The initial condition has the form

$$u(\mathbf{x}, 0) = f(\mathbf{x}), \quad \mathbf{x} \in D \cup \partial D. \tag{8.19}$$

Let a function $g = g(\mathbf{x}, t)$ (for instance continuous) be given on ∂D and the temperature distribution takes the prescribed values $g = g(\mathbf{x}, t)$ on the boundary:

$$u(\mathbf{x}, t) = g(\mathbf{x}, t), \quad \mathbf{x} \in \partial D. \tag{8.20}$$

The boundary condition (8.20) is called the *Dirichlet boundary value problem.*

The boundary condition can be prescribed as the normal flux on the boundary:

$$\mathbf{q}(\mathbf{x}, t) \cdot \mathbf{n}(\mathbf{x}) = h(\mathbf{x}, t), \quad \mathbf{x} \in \partial D, \tag{8.21}$$

where $h = h(\mathbf{x}, t)$ is given on ∂D. Using (8.3) we write the boundary condition (8.21) in the form called the *Neumann boundary value problem*

$$\nabla u(\mathbf{x}, t) \cdot \mathbf{n}(\mathbf{x}) = -\frac{h(\mathbf{x}, t)}{\lambda}, \quad \mathbf{x} \in \partial D. \tag{8.22}$$

It follows from (7.53) that (8.22) can be written as the boundary condition for the normal derivative

$$\frac{\partial u}{\partial \mathbf{n}}(\mathbf{x}, t) = -\frac{h(\mathbf{x}, t)}{\lambda}, \quad \mathbf{x} \in \partial D. \tag{8.23}$$

Let the boundary ∂D prevents the heat transfer, i.e., the body D is isolated. Then,

$$\frac{\partial u}{\partial \mathbf{n}}(\mathbf{x}, t) = 0, \quad \mathbf{x} \in \partial D. \tag{8.24}$$

The heat transfer between D and the surrounding medium can be more complicated. In accordance with Newton's law of cooling it can be modeled by the boundary condition

$$\lambda \frac{\partial u}{\partial \mathbf{n}}(\mathbf{x}, t) + \alpha[u(\mathbf{x}, t) - u_0(\mathbf{x}, t)] = 0, \quad \mathbf{x} \in \partial D, \tag{8.25}$$

where $u_0(\mathbf{x}, t)$ is an exterior temperature distribution on the boundary. The coefficient α characterizes the degree of ideal contact. If $\frac{\alpha}{\lambda} = 0$, we obtain the isolation condition (8.24). If $\frac{\alpha}{\lambda} = \infty$, we arrive at the Dirichlet condition $u(\mathbf{x}, t) = u_0(\mathbf{x}, t)$, i.e., at the perfect contact.

Composites are materials made from materials with different physical properties. Boundary (conjugation) conditions have to be posed on the joint boundary surface of two different phases. Let two materials of conductivities λ_1 and λ_2 have the prefect contact along a smooth surface S. The perfect contact means the equal temperatures and the normal fluxes from different sides of S, i.e.,

$$u_1 = u_2, \quad \mathbf{q}^{(1)} \cdot \mathbf{n} = \mathbf{q}^{(2)} \cdot \mathbf{n}, \quad \text{on} \quad S. \tag{8.26}$$

Here, u_i and $\mathbf{q}^{(i)} \cdot \mathbf{n}$ denote the limit values of the temperature and the normal heat flux on S from the side of the ith material ($i = 1, 2$). The second condition (8.26) can be written in the form

$$\lambda_1 \frac{\partial u_1}{\partial \mathbf{n}} = \lambda_2 \frac{\partial u_2}{\partial \mathbf{n}} \quad \text{on} \quad S. \tag{8.27}$$

The not-perfect contact between materials can be modeled by equations

$$\lambda_1 \frac{\partial u_1}{\partial \mathbf{n}} + \alpha[u_1 - u_2] = 0, \quad \lambda_1 \frac{\partial u_1}{\partial \mathbf{n}} = \lambda_2 \frac{\partial u_2}{\partial \mathbf{n}} \quad \text{na} \quad \partial D, \tag{8.28}$$

where the coefficient α characterizes the degree of the perfect contact as in the boundary condition (8.25).

8.3 Green's function for the 1D heat equation

Consider the 1D heat equation on the real axis with one spatial variable $x = x_1$

$$\frac{\partial u}{\partial t} = a^2 \frac{\partial^2 u}{\partial x^2}, \quad -\infty < x < \infty,\ t > 0. \tag{8.29}$$

Let the function $u(x,t)$ be replaced by the function $u(kx, k^2t)$ with a constant k. One can see that equation (8.29) does not change. Substitute $k = \frac{1}{2a\sqrt{t}}$ into $u(kx, k^2t)$ (one can use *Mathematica* as follows)

```
In[1]:= ∂t u[k x, k² t] - a² ∂x,x u[k x, k² t] // Simplify

Out[1]= k² (u^(0,1)[k x, k² t] - a² u^(2,0)[k x, k² t])

In[2]:= u[k x, k² t] /. k → 1/(2 a √t)

Out[2]= u[x/(2 a √t), 1/(4 a²)]
```

As a result we get $u(\frac{x}{2a\sqrt{t}}, \frac{1}{4a^2})$. One can note that it is worth introducing the variable $z = \frac{x}{2a\sqrt{t}}$ and the function

$$f(z) = u\left(\frac{x}{2a\sqrt{t}}, \frac{1}{4a^2}\right). \tag{8.30}$$

Substitution of (8.30) into (8.29) yields an ordinary differential equation. We can solve it and distinguish one of the solutions

```
In[3]:= ∂t f[x/(2 a √t)] - a² ∂x,x f[x/(2 a √t)] /. x → z 2 a √t // Simplify

Out[3]= -(2 z f'[z] + f''[z])/(4 t)

In[4]:= DSolve[2 z f'[z] + f''[z] == 0, f[z], z]

Out[4]= {{f[z] → C[2] + 1/2 √π C[1] Erf[z]}}

In[5]:= Simplify[Erf[z] /. z -> x/(2 a √t), Assumptions → a > 0]

Out[5]= Erf[x/(2 a √t)]
```

The function

$$\text{erf}(z) := \frac{2}{\pi}\int_0^z e^{-\xi^2}\, d\xi. \tag{8.31}$$

is called the *error function.* The integral (8.31) cannot be calculated in terms of elementary functions. But it is rather a psychological obstacle for users. Elementary functions constitute a class of "contractual" functions such as rational, power, trigonometric, logarithmic, exponential and their finite combinations including arithmetic operations and compositions. The symbol erf is just a mathematical encryption of such a function the derivative of which is equal to e^{-z^2}. So, the error function (8.31) is actually not more complicated than the elementary function $\sin z$. Using the principle by Dirac (stated on page 126) one can extend the class of elementary functions by addition of the error function but only after sufficiently extensive exercises with it.

Calculate the partial derivative

$$G(x,t) = \frac{1}{2}\frac{\partial}{\partial x}\left[\operatorname{erf}\left(\frac{x}{2a\sqrt{t}}\right)\right] = \frac{1}{2a\sqrt{\pi t}}e^{-\frac{x^2}{4a^2t}}. \tag{8.32}$$

It satisfies the heat equation (8.29):

```
In[6]:= G[x_, t_, a_] = 1/2 ∂x Erf[x/(2 a √t)]

Out[6]= e^(-x^2/(4 a^2 t)) / (2 a √π √t)

In[7]:= ∂t G[x, t, a] == a^2 ∂x,x G[x, t, a] // Simplify

Out[7]= True

In[8]:= G[x, t, a]

Out[8]= e^(-x^2/(4 a^2 t)) / (2 a √π √t)
```

Introduce the function $\Phi_m(x)$ (c.f. (6.45))

$$\Phi_m(x) = \frac{m}{\sqrt{\pi}}e^{-m^2x^2} \tag{8.33}$$

and put

$$m = \frac{1}{2a\sqrt{t}}. \tag{8.34}$$

Then,

$$\Phi_m(x) = G(x,t). \tag{8.35}$$

The limit of Φ_m as $m \to \infty$ was treated as the Dirac δ-function. It follows from (8.34) that $m \to \infty \Leftrightarrow t \to 0$. Hence, $G(x,t)$ tends to $\delta(x)$ as $t \to 0$. The function $\delta(x)$ models a heat source at the point $x = 0$ in stationary problems (see Sec.6.5). Therefore, the function $G(x,t)$ models a point instant heating of the unit heat intensity at the time $t = 0$ at the point $x = 0$. The bell shaped function $G(x,t)$ spreads in time to infinity (see Out[9] below and computer animation with the operator In[10]).

```
In[9]:= Plot[{G[x, 0.1, 1], G[x, 0.3, 1], G[x, 1, 1]}, {x, -2, 2},
  PlotStyle → Thick]
```

Out[9]=

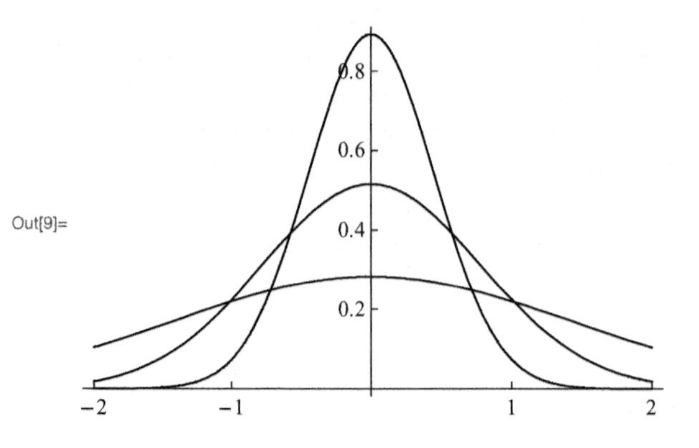


```
In[10]:= Animate[Plot[G[x, t, 1], {x, -1.5, 1.5}, PlotRange → {0, 5}],
    {t, .01, 0.1}]
```

The function $G(x,t)$ is called *Green's function* of the 1D heat equation on the line. Using $G(x,t)$ we can solve the initial problem $u(x,0) = f(x)$ for equation (8.29) on the real axis. Here, the boundary conditions go to infinity where the heat flux is bounded. Using the transition principle *continuous* $\leftrightarrow$ *discrete* we consider the temperature distribution $f(\xi_i)e^{-\frac{(x-\xi_i)^2}{4a^2t}}$ which arises due to a source located at the point $x = \xi_i$ at the initial time $t = 0$. Summation over all ξ_i and calculation of the limit of the obtained Riemann sum yield the desired temperature distribution in the integral form

$$u(x,t) = \frac{1}{2a\sqrt{\pi t}} \int_{-\infty}^{+\infty} f(\xi)e^{-\frac{(x-\xi)^2}{4a^2t}}\, d\xi. \qquad (8.36)$$

Similar formulae can be deduced for 2D and 3D problems when the temperature distribution is radially symmetric, i.e., $u = u(r,t)$, where $r = \sqrt{x_1^2 + x_2^2 + x_3^2}$.

Remark 8.2. In Sec.5.4, we discussed Brownian motion having used the discrete diffusion. Considering (8.29) as the diffusion equation with the diffusion constant $D = a^2$ we deduce formula (5.25) following Einstein (1905). Consider the function (8.32) which models the concentration density of the solute substance at time t. The total value of the substance (mass) is normalized to unity, i.e., $\int_{-\infty}^{+\infty} G(x,t)\, dx = 1$. By analogy with the heat conduction the unit mass of the substance is concentrated at the point $x = 0$ for $t = 0$ and changes in time as displayed in the above figure. The concentration density $G(x,t)$ can be treated as a probability density function of a continuous random variable X. The integral $\int_{x_1}^{x_2} G(x,t)\, dx$ is equal to the mass of substance on the interval (x_1, x_2) and to the probability of the event that the random variable X belongs to (x_1, x_2). Calculate the variance of X

$$Var[X] = \int_{-\infty}^{+\infty} x^2 G(x,t)\, dx = 2Dt. \qquad (8.37)$$

If we consider a solute like sugar suspended in tea, (8.37) shows how far the solute is spread out around the initial location $x = 0$. If we consider one particle, (8.37) estimates its distance to $x = 0$. This approach connects the deterministic and probabilistic worlds.

8.4 Fourier series

In the present section, we outlined constructive methods of the Fourier series applied in the next section to boundary value problems.

Let a function $f(x)$ be periodic with a period T. This function may be discontinuous at a finite number of points per period where it is left- and right- continuous. Moreover, let $f(x)$ be continuously differentiable at the intervals of continuity. These conditions can be weaken. The above conditions imply that the function $f(x)$ can be presented in the form of the convergent series

$$\frac{1}{2}[f(x+0)+f(x-0)] = \frac{a_0}{2} + \sum_{n=1}^{\infty}\left[a_n \cos\frac{2\pi nx}{T} + b_n \sin\frac{2\pi nx}{T}\right]. \quad (8.38)$$

The trigonometric series at the right side of (8.38) is called the *Fourier series.* Of course, if $f(x)$ is continuous at x, the left side of (8.38) is equal to $f(x)$ and (8.38) becomes

$$f(x) = \frac{a_0}{2} + \sum_{n=1}^{\infty}\left[a_n \cos\frac{2\pi nx}{T} + b_n \sin\frac{2\pi nx}{T}\right]. \quad (8.39)$$

The Fourier coefficients a_k and b_k are calculated in terms of $f(x)$

$$a_n = \frac{2}{T}\int_0^T f(x)\cos\frac{2\pi nx}{T}\,dx, \quad b_n = \frac{2}{T}\int_0^T f(x)\sin\frac{2\pi nx}{T}\,dx. \quad (8.40)$$

The complex form of the Fourier series can also be used

$$\frac{1}{2}[f(x+0)+f(x-0)] = \sum_{n=-\infty}^{+\infty} c_n \exp\left[\frac{2\pi nx}{T}\right], \quad (8.41)$$

where

$$c_n = \frac{1}{T}\int_0^T f(x)\exp\left[\frac{-2\pi nx}{T}\right]. \quad (8.42)$$

The coefficients a_n, b_n and c_n are related by formulae

$$c_n = \frac{1}{2}(a_n - ib_n), \quad c_{-n} = \frac{1}{2}(a_n + ib_n). \quad (8.43)$$

The complex Fourier series (8.41) represents a real function if $c_n = \overline{c_{-n}}$.

Parseval's identity takes place in the theory of Fourier series

$$\frac{1}{2}a_0^2 + \sum_{n=1}^{+\infty}(|a_n|^2 + |b_n|^2) = \sum_{n=-\infty}^{+\infty}|c_n|^2 = \frac{1}{T}\int_0^T f^2(x)\,dx. \tag{8.44}$$

In the case $T = 2\pi$, the above formulae can be simplified

$$\frac{1}{2}[f(x+0) + f(x-0)] = \frac{a_0}{2} + \sum_{n=1}^{\infty}(a_n \cos kn + b_n \sin nx)\,, \tag{8.45}$$

$$a_n = \frac{1}{\pi}\int_{-\pi}^{\pi} f(x)\cos nx\,dx, \quad b_n = \frac{1}{\pi}\int_{-\pi}^{\pi} f(x)\sin nx\,dx. \tag{8.46}$$

It follows from the periodicity that integration in (8.46) can be performed over an arbitrary interval of the length 2π. For even functions on $(-\pi,\pi)$ we get $b_k = 0$; for odd functions $a_k = 0$.

Fourier series (8.38) (or (8.45)) can be treated as a representation of a function on the basis $\{1, \cos nx, \sin nx,\ n = 1, 2, \ldots\}$ (cf. use of the operator **Fit** in Sec.5.2). Sometimes, it is convenient to take the basis $\{1, \cos nx\}$ or $\{\sin nx\}$. This can be done by consideration of a function on the interval $(0,\pi)$ what is always possible by a linear change of the argument and further artificial even or odd continuation of the function onto $(-\pi, 0)$.

The operator **FourierSinCoefficient** and others are useful for computations.

Example 8.2. Let the function $f(x)$ be determined by formula

```
In[1]:= f[x_] := { -1  Sin[x] < 0
                 {  1     True
```

Its graph is given in Fig.8.2.

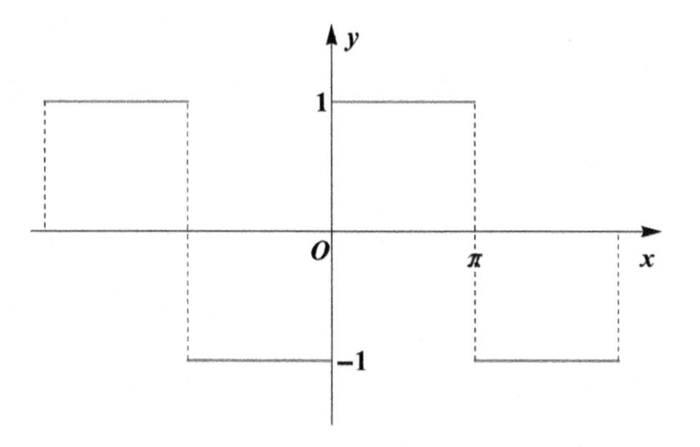

FIGURE 8.2: Graph of the function f.

This is an odd function, hence the cosine coefficients in its Fourier series vanish and only the sine coefficients have to be computed. The operator **FourierSinSeries** determines the series up to the desired order below, up to the twelve term.

```
In[2]:= FourierSinSeries[f[x], x, 12]
```

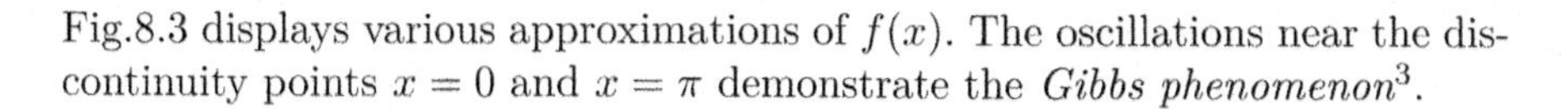

Let us define the nth approximation of the function $f(x)$:

```
In[3]:= f1[x_, n_] := FourierSinSeries[f[x], x, n]
```

Fig.8.3 displays various approximations of $f(x)$. The oscillations near the discontinuity points $x = 0$ and $x = \pi$ demonstrate the *Gibbs phenomenon*[3].

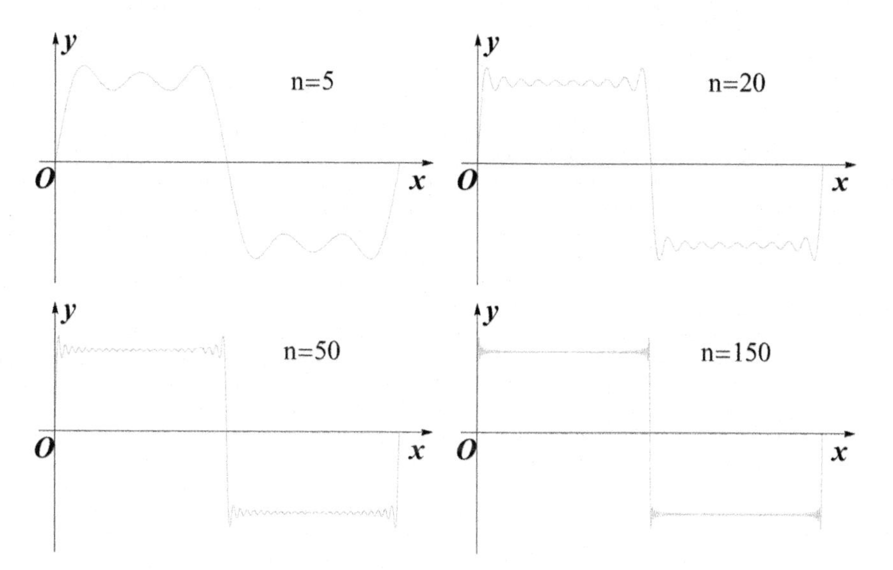

FIGURE 8.3: Graphs of the functions $y = f_1(x, n)$ dla $n = 5, 20, 50, 150$.

MATLAB Example Box 8.1

In the present box we create the animation of the approximation process of the *sawtooth* function:

$$f(x) = \frac{x}{\pi} - 2floor\left(\frac{1}{2} + \frac{x}{2\pi}\right).$$

[3]discovered by Henry Wilbraham (1848), see Arnold's Principle on page 159.

Again, the function is an odd one and its sine coefficients have the following form:

$$b_n = \frac{2(-1)^{n+1}}{n\pi}, n \geq 1.$$

```
function script17()

    % compute first n sine coefficients
    n = 1:50;
    bn = 2*(-1).^(n+1) ./ (pi*n);

    % define k-th approximation of f
    fk = @(x, k) sum(bn(1:k) .* sin((1:k)*x));
    x = linspace(-2*pi, 2*pi, 400);

    % plot function f and hold the picture
    y = x/pi - 2*floor(0.5 + x/(2*pi));
    plot(x, y, '-b');
    grid on
    axis([-2*pi, 2*pi, -1.5, 1.5])
    hold on

    % plot the first approximation
    y = arrayfun(@(x)fk(x, 0), x);
    fk_plot = plot(x, y, '-r');

    % animation loop
    for k=n
        y = arrayfun(@(x)fk(x, k), x);
        delete(fk_plot);
        fk_plot = plot(x, y, '-r');
        title(sprintf('f(x, k={%d})', k));
        pause(0.15); % wait some time
    end
    hold off
```

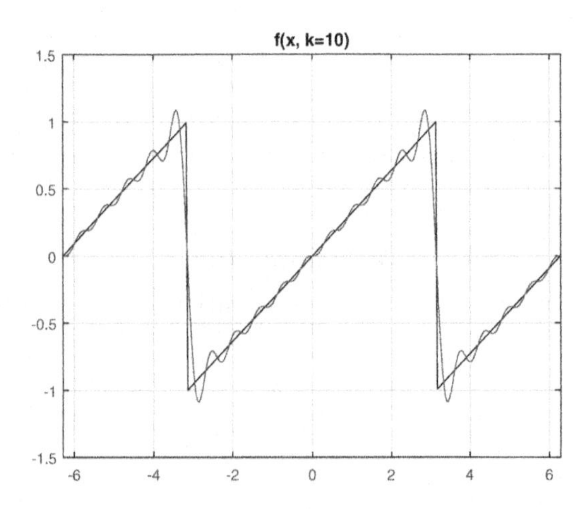

The animation is realized by simply updating some elements of the figure in `for` loop. While the `hold` is active, the plot of the current approximation is deleted and replaced with the next one. Note that the plot of the sawtooth function itself, as well as configuration of the figure, remain unchanged during the animation. The `arrayfun` operator above applies a function to each element of an array.

8.5 Separation of variables

Separation of variables is one of the most effective methods applied to ordinary and partial differential equations. The main idea is based on the reduction of an equation to such a form that its left and right sides contain different variables. If we get an equality $f(x) = g(t)$ where $f(x)$ depends only on the variable x and $g(t)$ depends only on the variable t, we conclude that the value

$$f(x) = g(t) \tag{8.47}$$

does not depend on x and t, hence it determines a constant λ for which $f(x) = g(t) = \lambda$. Therefore, we arrive at two separate equations $f(x) = \lambda$ and $g(t) = \lambda$ with one parameter λ.

The following differential equation can be solved by separation of variables

$$\frac{dx}{dt} = F(x)G(t) \tag{8.48}$$

by the scheme

$$\frac{dx}{F(x)} = G(t)\ dt \Leftrightarrow \int \frac{dx}{F(x)} = \int G(t)\ dt + c, \tag{8.49}$$

where $\int$ denotes one of the primitive functions and c is an arbitrary constant. After calculation of the integrals we arrive at an equation of the type (8.47) which determines an implicit function $x = x(t)$. Frequently, it is convenient to stop on equation (8.47) and to assume that two variables x and t are related by (8.47) without qualification of what is the argument and what is the function. Sec.4.2 contains an example of the above scheme.

In order to explain the main steps of the method for PDE we consider the 1D heat equation

$$\frac{\partial u}{\partial t} = a^2 \frac{\partial^2 u}{\partial x^2}, \quad 0 < x < \ell,\ t > 0, \tag{8.50}$$

with the initial condition

$$u(x, 0) = \phi(x), \quad 0 < x < \ell, \tag{8.51}$$

where a given function $\phi(x)$ is continuous in $0 \leq x \leq \ell$. Let the function $u(x,t)$ satisfy the boundary conditions

$$u(0,t) = u(\ell,t) = 0, \quad t > 0. \tag{8.52}$$

First, we are looking for $u(x,t)$ in the form

$$u(x,t) = X(x)T(t), \tag{8.53}$$

where $X(x)$ and $T(t)$ are functions of single variables. Substitute (8.53) into (8.50) (the arguments are omitted)

$$XT' = a^2X''T \tag{8.54}$$

Separate the variables dividing by XT

$$\frac{1}{a^2}\frac{T'}{T} = \frac{X''}{X} = -\lambda. \tag{8.55}$$

Here, the constant a^2 is put in the left side so that it does not impact onto the final result. The obtained function does not depend on x (left side) and on t (right side of the first equality). Hence, it is a constant taken as $-\lambda$ where the minus sign does not affect the final result. We have two separate ODE

$$X'' + \lambda X = 0 \tag{8.56}$$

and

$$T' + a^2\lambda T = 0. \tag{8.57}$$

The boundary conditions (8.52) for $X(x)$ become

$$X(0) = X(\ell) = 0. \tag{8.58}$$

Equations (8.56), (8.58) can be considered as the *spectral problem* [55] in which we find constants λ and the corresponding functions $X(x)$. The introduction of the whole problem into *Mathematica* gives only the trivial result:

```
In[1]:= DSolve[{X''[x] + λ X[x] == 0, X[0] == 0, X[1] == 0}, X[x], x]

Out[1]= {{X[x] → 0}}
```

Here, *Mathematica* follows the principle of the stupid computer (see page 10) and tries to solve the problem for any λ. Try to set *Mathematica* on the right track.

```
In[2]:= DSolve[{X''[x] + λ X[x] == 0, X[0] == 0}, X[x], x]

Out[2]= {{X[x] → C[2] Sin[x √λ]}}
```

```
In[3]:= DSolve[{X''[x] + λ X[x] == 0, X[1] == 0}, X[x], x]

Out[3]= {{X[x] → C[2] Sin[x √λ] - C[2] Cos[x √λ] Tan[1 √λ]}}
```

```
In[4]:= Simplify[Reduce[Sin[l √λ] == 0, λ], Assumptions → l > 0 && λ > 0]

Out[4]= C[1] ∈ Integers && ((√(C[1]^2) == C[1] && l^2 λ == 4 π^2 C[1]^2) ||
   (√((1 + 2 C[1])^2) == 1 + 2 C[1] && l^2 λ == (π + 2 π C[1])^2))
```

It looks much better. Even the result Out[2] is sufficient to find λ from equation

$$\sin\sqrt{\lambda}\ell = 0 \Leftrightarrow \lambda = \lambda_n := \left(\frac{\pi n}{\ell}\right)^2, \quad n = 1, 2, \ldots. \tag{8.59}$$

Therefore, the functions

$$X_n(x) = \sin\sqrt{\lambda_n}x \tag{8.60}$$

satisfy the problem (8.56), (8.58). Equation (8.57) with the corresponding λ has the solutions

$$T_n(t) = \exp(-a^2\lambda_n t), \quad n = 1, 2, \ldots. \tag{8.61}$$

Then, (8.53) becomes

$$u_n(x,t) = \exp(-a^2\lambda_n t)\sin\sqrt{\lambda_n}x, \quad n = 1, 2, \ldots. \tag{8.62}$$

A linear combination of (8.62)

$$u(x,t) = \sum_{n=1}^{\infty} U_n \exp(-a^2\lambda_n t)\sin\sqrt{\lambda_n}x \tag{8.63}$$

satisfies the differential equation (8.50). Here, U_n are undetermined constants. In order to construct the solution of the stated problem (8.50)-(8.52) we have to fulfil the initial condition (8.51) having at our disposal the coefficients U_n. Substitute $t = 0$ into (8.63)

$$u(x,0) = \sum_{n=1}^{\infty} U_n \sin\frac{\pi n}{\ell}x. \tag{8.64}$$

Represent the known function $\phi(x)$ from (8.51) in the form of the sinus Fourier series in the interval $(0, \ell)$

$$\phi(x) = \sum_{n=1}^{\infty} \phi_n \sin\frac{\pi n}{\ell}x, \tag{8.65}$$

where

$$\phi_n = \frac{2}{\ell}\int_0^{\ell} \phi(\xi)\sin\frac{\pi n}{\ell}\xi \, d\xi, \tag{8.66}$$

The uniqueness of the representation of a function by its Fourier series implies that the coefficients must coincide, $U_n = \phi_n$. This yields the final formula (8.63) with $U_n = \phi_n$.

Example 8.3. Let $\phi(x) = x$ and $\ell = 2\pi$. The coefficients ϕ_n are calculated as follows

```
In[1]:= ϕ[x_] := x

In[2]:= FourierSinCoefficient[ϕ[x], x, n, FourierParameters → {1, 1/2}]

Out[2]= -4 (-1)^n / n
```

The same result In[2] can be also obtained with (8.65):

```
In[3]:= ϕn = Simplify[1/π ∫_0^(2π) ϕ[ξ] Sin[n/2 ξ] dξ, Assumptions → n ∈ Integers]

Out[3]= -4 (-1)^n / n
```

The temperature distribution has the form

```
In[4]:= u[x_, t_] = Sum[% e^(-a λn t) Sin[π n/(2 π) x], {n, 1, ∞}]

Out[4]= Sum[-4 (-1)^n e^(-a t λn) Sin[n x/2] / n, {n, 1, ∞}]
```

Here, % in In[4] means the result obtained in the previous output, i.e., in Out[3]. One can substitute λ_n from (8.59) into In[4] and perform computations for numerical values of e and ℓ.

8.6 Discrete approximations of PDE

8.6.1 Finite-difference method

Boundary value problems stated in Sec.8.2 can be transformed into discrete problems following the principle of transition *continuous* $\leftrightarrow$ *discrete* (see page 19). As an example, consider the heat equation (c.f. (8.50)-(8.52))

$$\frac{\partial u}{\partial t} = a^2 \frac{\partial^2 u}{\partial x^2}, \quad 0 < x < \ell, \; t > 0, \tag{8.67}$$

with the initial condition

$$u(x, 0) = \phi(x), \quad 0 < x < \ell, \tag{8.68}$$

and the boundary conditions

$$u(0, t) = u_0(t), \quad u(\ell, t) = u_1(t), \quad t > 0, \tag{8.69}$$

where $u_0(t)$ and $u_1(t)$ are given boundary temperatures.

Below, we outline the finite-difference method effective in the numerical solution to the heat and other evolution equations[4]. Consider a rectangle $D = \{(x,t) : 0 < x < \ell, 0 < t < T\}$. Divide the segment $(0,\ell)$ on the x-axis onto small segments (x_i, x_{i+1}) $(i = 0, 1, \ldots, N-1)$. For definiteness, we take equal small segments putting $x_i = ih$ where h is called the spatial discretization step. Similar temporal discretization is introduced on the t-axis as $t_j = j\tau$ $(j = 0, 1, \ldots, M-1)$ with the discretization step τ. Instead of the continuous unknown function $u(x,t)$ we determine its values $u_{ij} = u(x_i, t_j)$ at the *mesh points* (x_i, t_j) by means of the discrete approximation of derivatives

$$\frac{\partial u}{\partial t} \approx \frac{1}{\tau}[u(x, t+\tau) - u(x,t)], \tag{8.70}$$

$$\frac{\partial^2 u}{\partial x^2} \approx \frac{1}{h^2}[u(x+h,t) - 2u(x,t) + u(x-h,t)]. \tag{8.71}$$

Formula (8.71) is based on the different approximations of the first derivative

$$\frac{\partial u}{\partial x}(x,t) \approx \frac{1}{h}[u(x+h,t) - u(x,t)] \text{ and } \frac{\partial u}{\partial x}(x-h,t) \approx \frac{1}{h}[u(x,t) - u(x-h,t)]. \tag{8.72}$$

Substitution of (8.70)-(8.71) into (8.67) yields

$$u_{i,j+1} = su_{i-1,j} + (1-2s)u_{i,j} + su_{i+1,j}, \tag{8.73}$$

where $s = \tau h^{-2}$. One can note that equation (8.73) in the second time variable j represent an *explicit scheme*. More precisely, the values $u_{i,j}$ are computed by layers counted by j. The set of values $u_{i,0} = \phi(x_i)$ is known from the initial condition (8.68). The first layer set $\{u_{i,1}\}$ is calculated by (8.73) with $j = 0$. The second layer set $\{u_{i,2}\}$ can be found from (8.73) with $j = 1$ and so forth (see Fig. 8.4). A numerical scheme must be stable, i.e., give close results with small perturbation of data (c.f., Sec.1.3). It can be proved that the scheme (8.73) is stable if $s < \frac{1}{2}$ [35].

Let $0 \le \sigma \le 1$ and the linear operator P is introduced as follows

$$Pu_{i,j} = s(u_{i-1,j} - 2u_{i,j} + u_{i+1,j}). \tag{8.74}$$

Instead of (8.73) we consider the scheme

$$u_{i,j+1} - u_{i,j} = \sigma Pu_{i,j+1} + (1-\sigma)Pu_{i,j}. \tag{8.75}$$

If $\sigma = 0$, we arrive at the explicit scheme (8.73). If $\sigma = 1$ we get the *implicit scheme*

$$u_{i,j+1} - u_{i,j} = Pu_{i,j+1} \equiv s(u_{i-1,j+1} - 2u_{i,j+1} + u_{i+1,j+1}). \tag{8.76}$$

[4]Here, we mean the parabolic PDE of the type $u_t = \dot{\nabla}(f(t,u)\nabla u) + F(t,u,\nabla u)$.

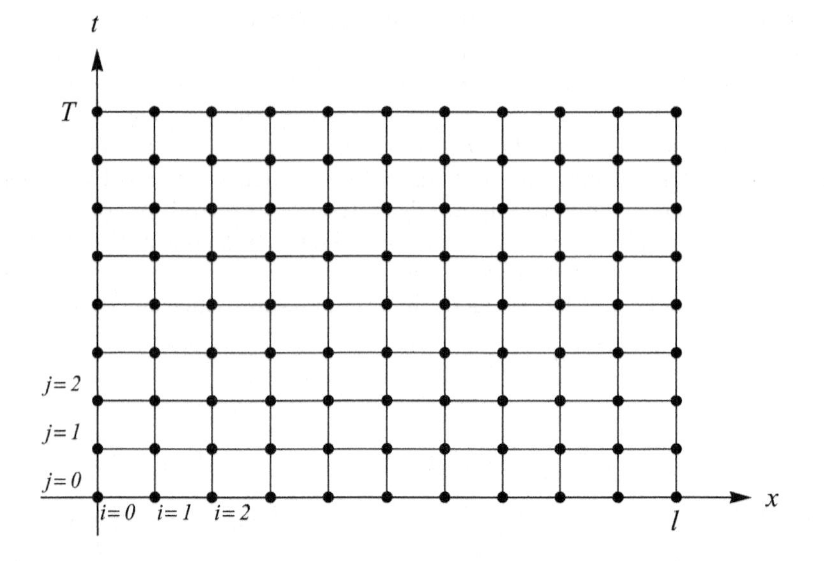

FIGURE 8.4: Mesh points with the double numeration (i, j); layers $j = 0$, $j = 1$, $j = 2$ etc.

The difference between (8.73) and (8.76) is in the shift on time $j \mapsto j+1$. As a result of this modification, in order to reach the layer $j = M-1$ we have to solve the system of linear algebraic equations (8.76) with $j = 1, 2, \ldots, M-1$ and $i = 1, 2, \ldots, N-2$. This system contains $(M-1)(N-2)$ equations on the same number of unknowns $u_{i,j}$. Therefore, the explicit scheme is more complicated than the explicit one in implementation. However, we have got the bonus that the explicit scheme is always stable[5]. Other values of σ can also be applied in simulations.

Sometimes *Mathematica* can solve PDE as it is illustrated in the following 1D example

```
In[1]:= NDSolve[{D[u[t, x], t] == D[u[t, x], {x, 2}], u[0, x] == 0,
      u[t, 0] ==
       Sin [t], u[t, 5] == 0}, u, {t, 0, 10}, {x, 0, 5}]

Out[1]= {{u → InterpolatingFunction[{{0., 10.}, {0., 5.}}, <>]}}
```

[5]We do not discuss here the numerical stability [35, 49] concerning the rounding error for large time-step sizes. However, the both types of stability has to be considered in practice. In this sense, the explicit scheme is stable ($s < \frac{1}{2}$) for a limited step size due to the accumulation of the error over each time step.

8.6.2 1D finite element method

Finite element methods are usually applied to stationary problems. In the present section, we outline one of these methods, the Rayleigh-Ritz method, to the Dirichlet problem for the simple second order differential equation in an interval $(a, b) \subset \mathbb{R}$

$$u'' = -f \quad \text{in } (a, b), \quad u(a) = u(b) = 0, \tag{8.77}$$

where $f(x)$ is a given function. Introduce the scalar product of two functions

$$[u, v] = \int_a^b u'v' \, dx. \tag{8.78}$$

This product induces a Hilbert space [16, 35] with the norm

$$\|u\| = \left(\int_a^b |u'|^2 \, dx \right)^{\frac{1}{2}}. \tag{8.79}$$

Integration by parts of functions $u(x)$ and $v(x)$ vanishing at the points a and b implies that

$$\int_a^b vu'' \, dx = \int_a^b v \, du' = -\int_a^b u' \, dv = -[u, v]. \tag{8.80}$$

Let u satisfies (8.77). Then, (8.80) yields

$$[u, v] = \int_a^b vf \, dx. \tag{8.81}$$

Divide the interval (a, b) into small intervals (x_{j-1}, x_j) where $j = 1, 2, \ldots, n+1$; $x_0 = a$ and $x_n = b$. Introduce the piece-linear functions called the basic functions for $j = 1, 2, \ldots, n$

$$u_j(x) = \begin{cases} \frac{x - x_{j-1}}{x_j - x_{j-1}} & x_{j-1} \le x \le x_j \\ \frac{x_{j+1} - x}{x_{j+1} - x_j} & x_j < x \le x_{j+1} \\ 0 & \text{otherwise} \end{cases} \tag{8.82}$$

The basic functions satisfies the relations $u_j(x_j) = 1$ and $u_j(x_{j+1}) = u_j(x_{j-1}) = 0$ (see Fig.8.5).

The next step is the approximation of the unknown function $u(x)$ by the linear combination of the basic functions

$$u(x) \approx \sum_{j=1}^{n} c_j u_j(x) \tag{8.83}$$

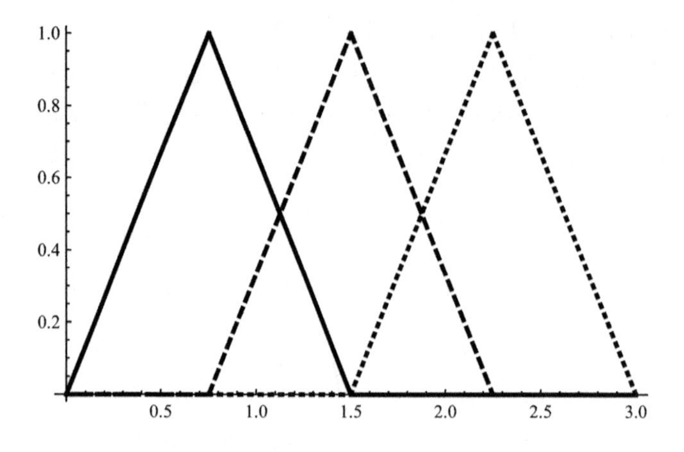

FIGURE 8.5: Plots of the basic functions.

with undetermined coefficients c_j. For each u_i introduce the linear operator $h \mapsto [u_i, h]$ which transforms a function h to the scalar product $[u_i, h]$ and apply it to equation (8.83). Using (8.81) we arrive at the system of linear algebraic equations on c_j

$$\sum_{j=1}^{n}[u_i, u_j]c_j = \int_a^b u_i f\, dx, \quad i = 1, 2, \ldots, n. \tag{8.84}$$

The coefficients $\{[u_i, u_j]\}$ form a symmetric matrix called the stiffness matrix.

MATLAB Example Box 8.2

Consider problem (8.77), where $f(x) = \sin(\pi x)$. One can see that the solution of the problem has the form

$$u(x) = \frac{\sin(\pi x)}{\pi^2}.$$

In order to find an approximation of $u(x)$, let us define basic functions (8.82), as well as their derivatives, and solve the system of linear algebraic equations (8.84) numerically.

```
function script18()

  a = 0;
  b = 3;
  n = 5;
  f = @(x) sin(pi*x);

  % points of the division of interval (a, b)
  x1 = linspace(a, b, n);
  % integral of product of two functions
```

```
    integr = @(u, v) integral(@(x) u(x).*v(x), a, b);

    % nested basic functions and their derivatives
    function y = u(j, x)
        y = zeros(size(x));
        int1 = (x >= x1(j-1)) & (x <= x1(j));
        int2 = (x > x1(j)) & (x <= x1(j+1));
        y(int1) = (x(int1)-x1(j-1))/(x1(j)-x1(j-1));
        y(int2) = (x1(j+1)-x(int2))/(x1(j+1)-x1(j));
    end
    function y = u_prime(j, x)
        y = zeros(size(x));
        int1 = (x >= x1(j-1)) & (x <= x1(j));
        int2 = (x > x1(j)) & (x <= x1(j+1));
        y(int1) =  1/(x1(j)-x1(j-1));
        y(int2) = -1/(x1(j+1)-x1(j));
    end

    % build matrices of the system of linear equations
    A = zeros(n-2,n-2);
    B = zeros(n-2,1);    % one-column matrix
    for i=1:n-2
        for j=1:n-2
            A(i, j) = integr(@(x) u_prime(i+1, x),...
                @(x) u_prime(j+1, x));
        end
        B(i) = integr(@(x) u(i+1, x), f);
    end
    % solve the system Ac=B
    c = A\B;

    % plot the solution and compare with f
    x = linspace(a, b, 100);
    y = zeros(size(x));
    for j=2:n-1
        y = y + c(j-1) * u(j, x);
    end
    plot(x, f(x)/pi^2, x, y)
    grid on
end
```

One can change the accuracy of the solution altering the number of division points n of the interval (a, b). Below, the cases of $n = 5$ and $n = 13$ are shown.

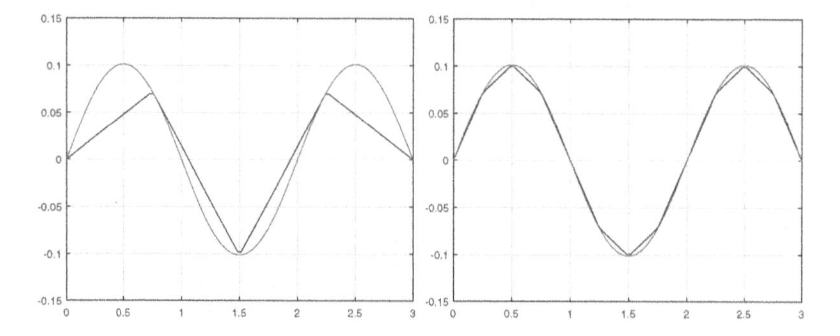

Note the definitions of the *nested* functions `u(j, x)` and `u_prime(j, x)` taking advantage of logical vectors in order to designate the intervals for the proper values. Both functions operate on arrays and have access to the variables from the workspace of the enclosing function (i.e. the variable `x1`).

In the above example, we solve the system of linear equations $Ax = B$ with B being a column vector. One can also solve the equivalent system $x^T A^T = B^T$, where T denotes the transposition operation:

```
>> xt = Bt/At
```

8.6.3 Finite element method in $\mathbb{R}^2$

In the present section, we apply the Rayleigh-Ritz method, to the Dirichlet problem for the Poisson equation in a bounded 2D domain D

$$\nabla^2 u = -f \quad \text{in } D, \quad u = 0 \quad \text{on } \partial D, \tag{8.85}$$

where f is a given function. One can compare this section line by line with the previous one devoted to the simple 1D case.

Introduce the scalar product of two functions

$$[u, v] = \int_D \nabla u \cdot \nabla v \, d\mathbf{x}. \tag{8.86}$$

This product induces a Hilbert space [16, 35] with the norm

$$\|u\| = \left(\int_D |\nabla u|^2 \, d\mathbf{x} \right)^{\frac{1}{2}}. \tag{8.87}$$

Application of Theorem 7.3 (Green's formula (7.88) on page 160) to functions u and v vanishing on the boundary ∂D implies that

$$\int_D v \nabla^2 u \, d\mathbf{x} = -[u, v]. \tag{8.88}$$

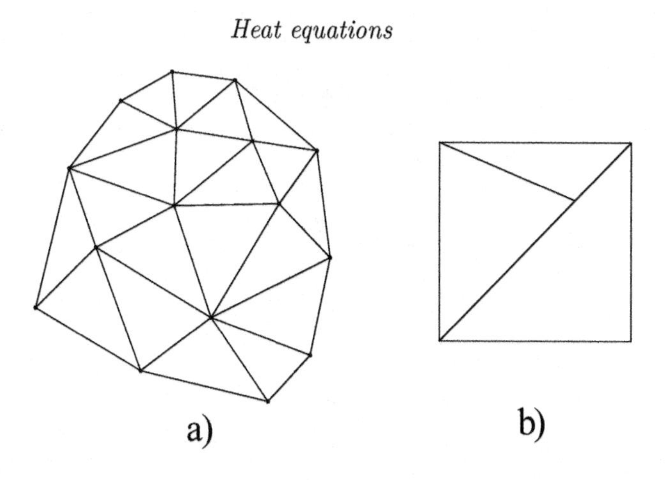

FIGURE 8.6: a) Proper triangulation; b) non-allowable triangulation.

Let u satisfy (8.85). Then, (8.88) yields

$$[u, v] = \int_D vf \, d\mathbf{x}. \tag{8.89}$$

A triangulation of the domain D means an approximation of D by the union of triangles. Not all sets of triangles fit for a triangulation. A typical allowable triangulation is shown in Fig.8.6. Let $\mathbf{v}_j$ be vertices of the triangulation $(j = 1, 2, \dots, n)$. A proper triangulation must satisfy the following condition. Piecewise linear functions constructed by assigning arbitrary values at $\mathbf{v}_j$ are continuous. This restriction allows us to construct a set of basic functions $u_j(\mathbf{x})$ $(j = 1, 2, \dots, n)$.

Fix a vertex $\mathbf{v}_j$ and consider all triangles Δ_m $(m = 1, 2, \dots, N_j)$ containing $\mathbf{v}_j$ as shown in Fig.8.7. Introduce a basic function $u_j(\mathbf{x}) = u_j(x_1, x_2)$ as a linear function $= a_0^{(jm)} + a_1^{(jm)} x_1 + a_2^{(jm)} x_2$ in each triangle Δ_m. The coefficients $a_k^{(jm)}$ are chosen in such a way that $u_j(\mathbf{v}_j) = 1$ and $u_j(\mathbf{v}_l) = 0$ for $l \neq j$. Such a choice is correctly defined since we have three conditions (at each vertex of triangle) on three coefficients $a_0^{(jm)}$, $a_1^{(jm)}$ and $a_2^{(jm)}$.

The next step is the approximation of the solution $u(\mathbf{x})$ of the problem (8.85)

$$u(\mathbf{x}) \approx \sum_{j=1}^{n} c_j u_j(\mathbf{x}) \tag{8.90}$$

with undetermined coefficients c_j. For each u_i introduce the linear operator $h \mapsto [u_i, h]$ and apply it to equation (8.90). Using (8.89) we arrive at the system of linear algebraic equations on c_j

$$\sum_{j=1}^{n} [u_i, u_j] c_j = \int_D u_i f \, d\mathbf{x}, \quad i = 1, 2, \dots, n. \tag{8.91}$$

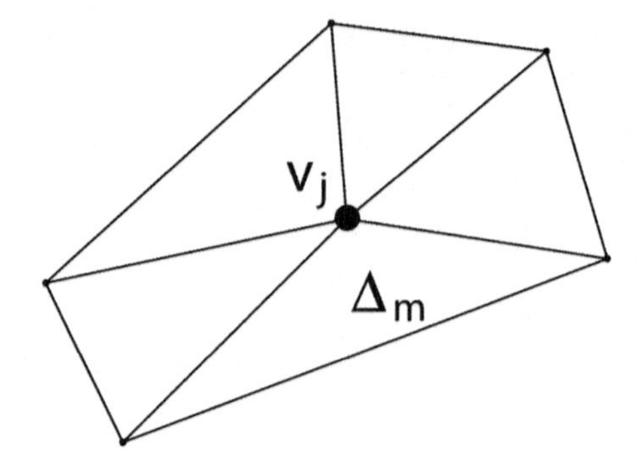

FIGURE 8.7: Basic function $u_j(\mathbf{x})$ is a linear function in each triangle Δ_m, equal to 1 at the central point $\mathbf{v}_j$ and 0 at other vertices.

The coefficients $\{[u_i, u_j]\}$ form the stiffness matrix.

Mathematica contains the finite element package which can be loaded by the operator `Needs["NDSolve`FEM`"]`. The 2D Dirichlet problem for the Laplace equation in a rectangle can be solved by use of the operator `NDSolveValue[{Laplacian[u[x, y], {x, y}] == 0, u[x, 0] == 0, u[x, 1] == 1, u[0, y] == 0, u[1, y] == 0}, u, {x, 0, 1}, {y, 0, 1}]`.

8.7 Universality in Mathematical Modeling. Table

TABLE 8.1: Universality in Mathematical Modeling [10]. Physical phenomena and corresponding vector fields $\mathbf{j}(\mathbf{x})$, $\mathbf{E}(\mathbf{x})$; tensors $\boldsymbol{\sigma}(\mathbf{x})$ and scalars for anisotropic and isotropic media, respectively. The fields satisfy equations $\mathbf{j}(\mathbf{x}) = \boldsymbol{\sigma}(\mathbf{x})\mathbf{e}(\mathbf{x})$ and $\nabla \cdot \mathbf{j} = 0$. In 1D case, the operator ∇ is the derivative, i.e., $\nabla = \frac{d}{dx}$. Then, $j(x)$, $E(x)$ and $\sigma(x)$ are scalar function in x. Equations $j(x) = \sigma(x)e(x)$ and $\frac{dj}{dx} = 0$ are fulfilled for 1D media.

Topics	$j(\mathbf{x})$	$e(\mathbf{x})$	$\boldsymbol{\sigma}(\mathbf{x})$
Electrical conduction	Electrical current $\mathbf{j}$	Electric field $\mathbf{e}$	Electrical conductivity σ
Dielectrics	Displacement field $\mathbf{d}$	Electric field $\mathbf{e}$	Electric permittivity ε
Magnetism	Magnetic induction $\mathbf{B}$	Magnetic field $\mathbf{H}$	Magnetic permeability μ
Heat conduction	Heat flux q	Temperature gradient $-\nabla T$	Thermal conductivity λ
Diffusion	Particle current j	Concentration gradient $-\nabla c$	Diffusivity D
Flow in porous media	Velocity v	Pressure gradient ∇P	Fluid permeability K
Antiplane elasticity	Stresses (τ_{13}, τ_{23})	Vertical displacement gradient ∇u_3	Shear modulus G

Further reading. We recommend the textbooks [41, 52, 55] and the advanced books [6, 11, 16, 22, 26, 31, 33, 36, 42].

Exercises

1. Present the a graph of the function erf x. Calculate its derivative and primitive.

2. Find the temperature distribution $u(x,t)$ governed by equation (8.29) with the initial condition

$$u(x,0) = \begin{cases} A & \text{for} \quad x_1 \leq x < x_2, \\ B & \text{otherwise} \end{cases}$$

   where $A = 1$ and $B = 0$.

   Present the graph of the function $u(x,t)$ using animation in t.

   Solve the problem for arbitrary A and B.

3. Let the pipe from Example 8.1 be made from glass and its radii be $r_1 = 2cm$ and $r_2 = 2.5cm$. Let the interior and exterior temperatures be $T_1 = 75^0$ and $T_2 = 25^0$. Determine the temperature distribution in the pipe.

4. Following Example 8.1 investigate the heat conduction of the spherical layer with the interior and exterior radii r_1 and r_2 filled by material of conductivity λ.

5. Using Example 8.1 and the linear temperature distribution for layered media from the first two sections of Chapter 6, compare the heat fields in the layer of the width ℓ, in the hollow cylinder and in the spherical layer of the width $\ell = r_2 - r_1$. The boundary temperatures are equal to T_1 and T_2.

   Consider the same problem when instead of the temperature the normal heat flux is given on the boundary. Show that the normal derivative to sphere coincides with the radial derivative $\frac{\partial}{\partial n} = \frac{\partial}{\partial r}$.

   Consider various combinations of the boundary temperature and flux.

6. Deduce 2D and 3D Green's functions similar to 1D function (8.32).

7. Display the distribution temperature from Example 8.3 in time using animation.

8. Represent the following functions by their Fourier series in $(-\pi, \pi)$ and by their cosine Fourier series in $(0, \pi)$:

   $\sin x$;

   $\sin 2x$;

   $\sin \frac{3}{2}x$;

   the Dirac δ-function.

9. Investigate the temperature distribution of the 1D layered composite when the first layer $0 < x < x_1$ is occupied by material of conductivity λ_1, the second layer $x_1 < x < x_2$ by material of conductivity λ_2.

The contact between layers is assumed to be perfect. Consider various combinations of the boundary temperature and flux.

Calculate the averaged heat flux

$$\langle q \rangle = -\frac{1}{x_2}\left(\int_0^{x_1} \lambda_1 u'(x)\mathrm{d}x + \int_{x_1}^{x_2} \lambda_2 u'(x)\mathrm{d}x\right)$$

when the normalized flux $q = 1$ is given (formally only on the boundary). Calculate the effective conductivity $\lambda_\perp$ perpendicular to the layers which is defined by the averaged Fourier law

$$\langle q \rangle = -\lambda_\perp (T_2 - T_1)/x.$$

Calculate the effective conductivity $\lambda_{||}$ along the layers.

Answer:

$$\lambda_\perp = \frac{\ell}{\frac{\ell_1}{\lambda_1} + \frac{\ell_2}{\lambda_2}} \quad \text{and} \quad \lambda_{||} = \frac{1}{\ell}(\ell_1\lambda_1 + \ell_2\lambda_2), \tag{8.92}$$

where $\ell_1 = x_1$, $\ell_2 = x_2 - x_1$, $\ell = x_2$.

Chapter 9

Asymptotic methods in composites

9.1 Effective properties of composites 194
9.1.1 General introduction 194
9.1.2 Strategy of investigations 196
9.2 Maxwell's approach 199
9.2.1 Single-inclusion problem 199
9.2.2 Self consistent approximation 201
9.3 Densely packed balls 202
9.3.1 Cubic array 202
9.3.2 Densely packed balls and Voronoi diagrams 204
9.3.3 Optimal random packing 208
Exercises 213

Asymptotic methods [3] are assigned to analytical methods when solutions are investigated near the critical values (singular points) of the geometrical and physical parameters. Hence, asymptotic formulae can be considered as analytical approximations. Asymptotic methods are more than a set of procedures. They represent an ideology of the qualitative study of phenomena [3].

Principle of Asymptotology *The key properties of the system (process, object) are manifested near singular points. These singular points determine the behavior of the system in regular regimes.*

A typical application of asymptotic methods can be shortly demonstrated by the following example. Consider an elastic stress problem for fractured media. The stress tensor $\sigma(\mathbf{x})$ is not bounded around the crack tip located at a point $\mathbf{x} = \mathbf{a}$. Numerical straightforward approximations of $\sigma(\mathbf{x})$ could be very expensive to reach a satisfactory result because any numerical approximation works for finite values. Asymptotic analysis, first, suggests to analytically investigate the local stress by means of the asymptotic formula $\sigma(\mathbf{x}) \sim \frac{K}{\sqrt{r}}$ where $r = |\mathbf{x} - \mathbf{a}|$, i.e., to estimate the constant K. The next step can be based on a numerical procedure applied to the bounded value $\sigma(\mathbf{x}) - \frac{K}{\sqrt{r}}$. Most commercial FEM software packages include such asymptotic formulae.

Actually, we applied the principle of Asymptotology in Sec.4.8 to nonlinear differential equations (4.65), (4.67) having reduced them to the linear equations (4.69) by asymptotic approximations near the stationary point. In

Sec.4.5.2, we exploit asymptotic expansions and Padé approximations to approximate solutions of ODE near singular points. In the present chapter, we apply asymptotic methods to estimation of the effective properties of composites.

9.1 Effective properties of composites

9.1.1 General introduction

A composite is a material made from a few constituent materials with different physical properties. For definiteness, we consider the heat conduction characterized locally by the thermal conductivity $\lambda(\mathbf{x})$. We restrict our attention to two-component composites made from a collection of non-overlapping, identical, conducting balls of radius r_0, embedded in an otherwise uniform locally isotropic host. Let λ_1 and λ denote the conductivity of spherical inclusions and of the host, respectively. The local Fourier law (8.3) in the considered case becomes

$$\mathbf{q} = -\lambda(\mathbf{x})\nabla u = \begin{cases} -\lambda_1 \nabla u, & \text{if } \mathbf{x} \text{ belongs to inclusions,} \\ \\ -\lambda \nabla u, & \text{if } \mathbf{x} \text{ belongs to host.} \end{cases} \tag{9.1}$$

Consider a sample of a macroscopically isotropic composite containing a sufficiently large number N of non-overlapping spherical inclusions displayed in Fig.9.1. An experimentator can treat this sample as a homogeneous material and determine its conductivity $\widehat{\lambda}$ with (6.2) by measure of the normal flux on one of the faces of the sample and of the difference of the constant temperatures between this and the opposite faces.

In order to state the corresponding computational experiment we consider the sample as a domain $G = D \cup H \cup \partial H$ occupied by the composite, where $H = \cup_{k=1}^{N} D_k$ denotes the union of all balls, ∂H the union of all spheres, and D the host domain. The concentration of inclusions ν is introduced as the ratio of volumes $\nu = \frac{|H|}{|G|}$.

The proper choice of the sample and its size is a separate task. The properly constructed sample of minimally possible size is called the representative volume element (RVE). One can find a method to determine RVE in [53, 26].

It is convenient to determine $\widehat{\lambda}$ by the averaged value over G

$$\langle f(\mathbf{x}) \rangle = \frac{1}{|G|} \int_G f(\mathbf{x})\, d\mathbf{x}. \tag{9.2}$$

Let $\langle \mathbf{q} \rangle$ denote the averaged flux and $\widehat{\nabla u}$ the macroscopic temperature gradient over the considered sample. The effective conductivity $\widehat{\lambda}$ in the case of the

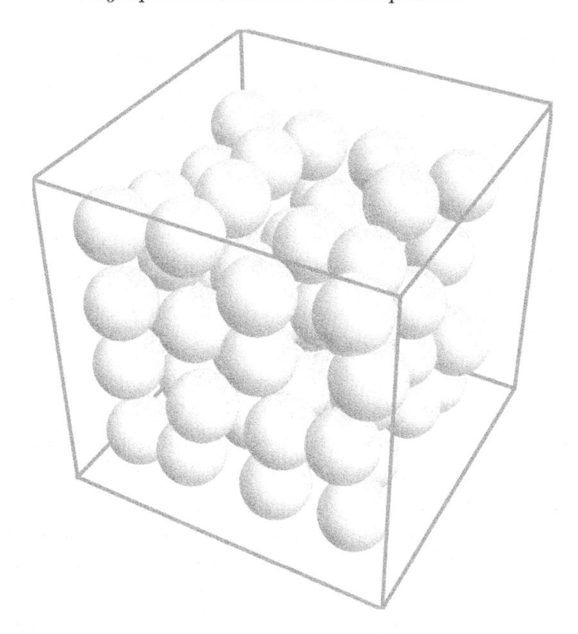

FIGURE 9.1: Spherical inclusions embedded in host.

perfect contact between inclusions and host is defined as the proportionality coefficient in equation [26]

$$\langle \mathbf{q} \rangle = -\widehat{\lambda}\widehat{\nabla u}. \tag{9.3}$$

This approach is similar to the black box operator discussed in Sec.1.2 (page 20). Let the input $\widehat{\nabla u}$ be given and the output $\langle \mathbf{q} \rangle$ be measured (computed). Then, the coefficient $\widehat{\lambda}$ determines the conductive properties of the sample G, regardless of its microstructure. In the same time, microstructure determines $\widehat{\lambda}$. Therefore, in the computer simulations as well as in real experiments we have to take the same statistically equivalent composites [26].

The linear equation (9.3) relates two constant vectors $\langle \mathbf{q} \rangle$ and $\widehat{\nabla u}$ in $\mathbb{R}^3$. It follows from algebra that a linear transformation of vectors is determined by a matrix called the effective conductivity tensor (see Sec.7.5)

$$\widehat{\boldsymbol{\lambda}} = \begin{pmatrix} \lambda_{11} & \lambda_{12} & \lambda_{13} \\ \lambda_{21} & \lambda_{22} & \lambda_{23} \\ \lambda_{31} & \lambda_{32} & \lambda_{33} \end{pmatrix}. \tag{9.4}$$

The tensor (9.4) is symmetric, hence, it can be reduced to the diagonal form as noted at the end of Sec.7.5. The tensor $\widehat{\boldsymbol{\lambda}}$ express the anisotropic effective properties of the homogenized material, i.e. different conductivity in different directions. Let the matrix (9.4) be represented in the form $\widehat{\boldsymbol{\lambda}} = \widehat{\lambda} I$, where I denotes the unit matrix. In this case, the considered composite is called macroscopically isotropic and the scalar $\widehat{\lambda}$ is called its effective conductivity.

The constructive theory of composites is presented in [26] and works cited therein. We shall shortly present its methods in academic form except in the next section where we outline eventual first steps by a beginning applied mathematician studying composites.

9.1.2 Strategy of investigations

Let an engineer-user ask us, the applied mathematicians, to estimate the effective conductivity of a composite consisting of spherical particles embedded in the host as displayed in Fig.9.1. It is assumed that the considered composite is macroscopically isotropic. Let us have the data at our disposal, the conductivities of components λ_1, λ and the concentration ν. Following Sec.1.1.3 we have to develop a mathematical model. This means, first, that we have to properly select a model from a set of models scattered everywhere, c.f. internet and textbooks.

As an example, we present the following scenario of such a work. Let us find two formulae for the effective conductivity. The first formula is called the rule of mixtures

$$\widehat{\lambda} \approx \frac{1}{|G|} \int_G \lambda(\mathbf{x})\, d\mathbf{x} = \nu\lambda_1 + (1-\nu)\lambda. \tag{9.5}$$

The second formula is called the Clausius-Mossotti approximation

$$\frac{\widehat{\lambda}}{\lambda} \approx \frac{1+2\varrho\nu}{1-\varrho\nu}. \tag{9.6}$$

Here, the so-called contrast parameter is introduced

$$\varrho = \frac{\lambda_1 - \lambda}{\lambda_1 + 2\lambda} = \frac{\frac{\lambda_1}{\lambda} - 1}{\frac{\lambda_1}{\lambda} + 2}. \tag{9.7}$$

It is equal to zero if $\lambda_1 = \lambda$, i.e., the inclusions merge with the host. If $\frac{\lambda_1}{\lambda} = 0$ $\Leftrightarrow \varrho = -\frac{1}{2}$, the inclusions become empty holes (isolators). If $\frac{\lambda_1}{\lambda} = \infty \Leftrightarrow \varrho = 1$, the inclusions become ideal conductors.

Formula (9.5) looks sensible but its careful study suggests that it is valid rather for layered composites, not for dispersed media. One can note that (9.5) coincides with the conductivity $\lambda_{||}$ along layers discussed on page 191.

Formula (9.7) looks more suitable since it refers to spherical inclusion diluted in the host. After discussions with the engineer-user we can stop at (9.7) and finish the work. This is an optimistic scenario.

The pessimistic scenario can be realized if we stumble across the webpage displayed in Fig.9.2. It demonstrates the methodologically wrong scheme of the RVE frequently used by engineer-users in practice when a random composite is replaced by a regular one. This misleading scheme implies a numerical solution to the problems for the unit cells depicted in the third column of Fig.9.2.

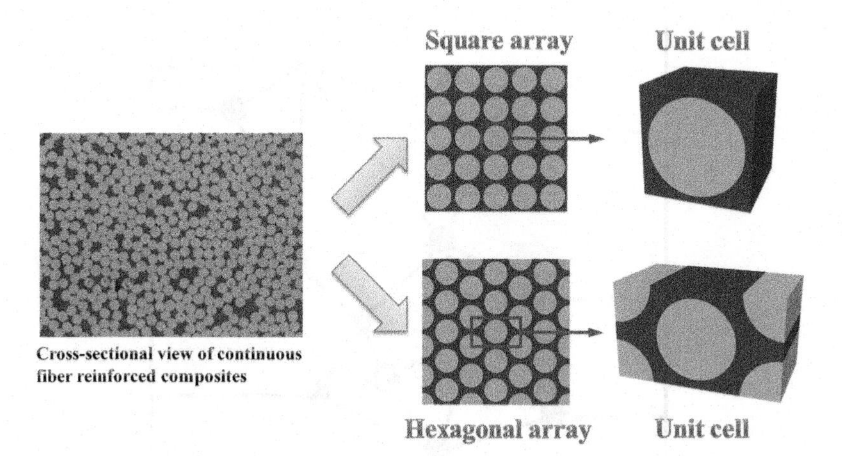

FIGURE 9.2: Methodologically wrong scheme of the RVE from WikipediA *en.wikipedia.org/wiki/Representative_elementary_volume*. The random (on the left) and regular (on the right) composites in Fig.9.2 can have significantly different macroscopic properties in the case of percolation displayed in Fig.9.5.

Moreover, the double periodicity cell for the hexagonal array is presented as a unit cell that doubles the discretization domain. Let one select this scheme adopted to 3D composites and perform hard work. The result can surprise, since in many cases it will be close to (9.6) because the approximation (9.6) is valid for dilute composites when $|\varrho\nu|$ does not exceed 0.3 for conductivity problems[1]. After verification of the model (see Sec.1.1.3) we can come to the conclusion that we need other formulae for highly dense structures.

We can stumble across the book [26] where many useful analytical formulae can be found. Random structures will shortly be discussed in Sec.9.3 by advanced asymptotic methods. Now, we summarize a graph computational scheme using a 2D scheme.

Let the centers of inclusions $\mathbf{x}_k$ $(k = 1, 2, \cdots, N)$ belong to the unit cube $C = \{\mathbf{x} \in \mathbb{R}^3 : |x_i| < \frac{1}{2},\ i = 1, 2, 3\}$. Let the points form a Delaunay graph (see Fig. 9.3 and description in Sec.9.3).

The following discrete model of the heat conduction through the cube (square) is considered [11, 36]. Let the temperature on the vertical faces of the cube be equal to 0 on $x_1 = -\frac{1}{2}$ and 1 on $x_1 = \frac{1}{2}$, and the unknown temperature at $\mathbf{x}_k$ $(k = 1, 2, \ldots, N)$ be denoted by u_k. The relative interparticle flux g_{km} between neighbor points $\mathbf{x}_k$ and $\mathbf{x}_m$ is estimated by formula (see justification of (9.37) below for periodic media)

$$g_{mk} = \lambda\pi r_0 \ln \frac{r_0}{\|\mathbf{x}_m - \mathbf{x}_k\| - 2r_0}, \tag{9.8}$$

where r_0 denotes the radius of spherical inclusions. It is convenient to intro-

[1] The dilute case usually means that $\nu \leq 0.05$ for elastic composites and suspensions.

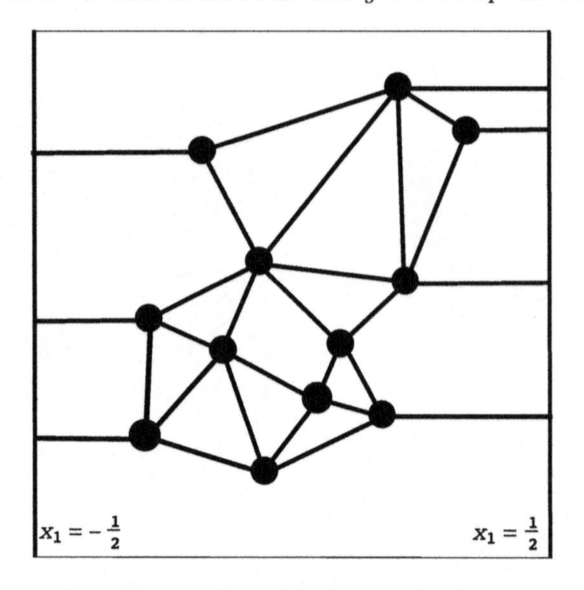

FIGURE 9.3: 2D interpretation.: square with random points forms a Delaunay graph.

duce the mega-vertices $\mathbf{x}_0$ and $\mathbf{x}_{N+1}$ corresponding to the sides $x_1 = -\frac{1}{2}$ and $x_1 = \frac{1}{2}$, respectively. Let $j \sim k$ means that vertices $\mathbf{x}_k$ and $\mathbf{x}_j$ are connected. Then, the set of u_k satisfies the system of linear algebraic equations

$$\sum_{j\sim k} g_{kj}(u_k - u_j) = 0, \quad k = 1, 2, \ldots, N. \tag{9.9}$$

The system (9.9) contains N unknowns u_k $(k = 1, 2, \ldots, N)$ and two parameters $u_0 = 0$, $u_{N+1} = 1$. It is possible to prove that the system (9.9) has a unique solution in non-degenerated cases. The effective conductivity of the sample is calculated by formula

$$\widehat{\lambda} = \frac{1}{2}\sum_{k=0}^{N+1}\sum_{j\sim k} g_{kj}(u_k - u_j)^2. \tag{9.10}$$

One can check that the minimization of the function (9.10) on the variables u_k $(k = 1, 2, \ldots, N)$ yields the system (9.9). This minimization coincides with the method of least squares discussed in Sec.5.1.

One can go ahead and minimize the value (9.10) on the geometric parameters $\mathbf{x}_k$ $(k = 1, 2, \ldots, N)$. Such an approach can be applied to investigation of pattern formation usually described by complicated reaction-diffusion equations [22, 33, 48]. The considered discrete approach explains formation of regular structures in chemistry, biology etc. In particular, it explains why hexagonal arrays are more frequently met in nature than square arrays, for instance, why the leopard skin has a structure reminiscent of the disturbed hexagonal array.

9.2 Maxwell's approach

Following Maxwell [45] we evaluate the effective conductivity $\widehat{\lambda}$ of dilute composites with spherical inclusions when concentration ν is sufficiently small. Maxwell's asymptotic approach is based on the assumption that inclusions are far away from each other and the interactions among inclusions are not taken into account. Maxwell's approach consists of two steps. First, a problem for one inclusion is solved and the so-called perturbation term (dipole) at infinity is found. Second, the homogenized medium is introduced and its perturbation term is equated to the sum of disturbance terms for many balls. The obtained equation determines the effective conductivity of dilute dispersed composites.

9.2.1 Single-inclusion problem

In order to determine the local field near each inclusion Maxwell considered the single-inclusion problem when the temperature distribution $u(\mathbf{x})$ fulfils the Laplace equation (8.14) both inside and outside the sphere $|\mathbf{x}| = r_0$ except infinity where

$$u(\mathbf{x}) \sim x_1. \tag{9.11}$$

The latter condition means that the plane-parallel flux

$$\mathbf{q}_\infty = -\lambda \nabla x_1 = -\lambda(1,0,0) \tag{9.12}$$

is applied at infinity. Following Sec.8.1 (see (8.27)) the perfect contact between the components is expressed by the relations

$$u_1 = u, \quad \lambda_1 \frac{\partial u_1}{\partial \mathbf{n}} = \lambda \frac{\partial u}{\partial \mathbf{n}} \quad \text{on} \quad |\mathbf{x}| = r_0, \tag{9.13}$$

where $u_1(\mathbf{x})$ and $u(\mathbf{x})$ stand for the temperature distribution in $|\mathbf{x}| < r_0$ and in $|\mathbf{x}| > r_0$, respectively. The normal derivative on the sphere $\frac{\partial}{\partial \mathbf{n}}$ is equal to $\frac{\partial}{\partial r}$, where $r = |\mathbf{x}|$ denotes one of the spherical coordinates discussed in Sec.7.1.3. Therefore, (9.13) can be written in the form

$$u_1 = u, \quad \lambda_1 \frac{\partial u_1}{\partial r} = \lambda \frac{\partial u}{\partial r} \quad \text{on} \quad |\mathbf{x}| = r_0. \tag{9.14}$$

One can suggest to look for $u_1(\mathbf{x})$ and $u(\mathbf{x})$ in the form of the scalar product (in the Cartesian coordinates)

$$U(\mathbf{x}) = (1,0,0) \cdot \nabla v(r) = \frac{x_1}{r} \frac{dv}{dr}(r), \tag{9.15}$$

where $v(r)$ is a function of one variable $r = \sqrt{x_1^2 + x_2^2 + x_3^2}$. Here, the chain rule $\frac{\partial v}{\partial x_1} = \frac{dv}{dr}\frac{\partial r}{\partial x_1}$ and formula based on (7.16)

$$\frac{\partial x_1}{\partial r} = \sin\theta\cos\phi = \frac{x_1}{r}. \tag{9.16}$$

are used. The form (9.15) is suggested by physical observations and by the radial symmetry. It is possible to explain formula (9.15) by the observation that the vector $(1,0,0)$ is associated to the external flux (9.12). Anyway, the suggestion (9.15) will lead to the result. Actually, one can try a few reasonable attempts and hit the target (9.15) according to the principle of trying presented on page 124.

The function (9.15) satisfies the Laplace equation, hence

$$\nabla^2 U(\mathbf{x}) = (1,0,0)\cdot\nabla(\nabla^2 v(r)) = 0. \tag{9.17}$$

This implies that $\nabla^2 v(r)$ is a constant. Using the form of the Laplace operator in spherical coordinates (see page 158) we obtain the differential equation

$$\frac{d}{dr}\left(\frac{2}{r}\frac{dv}{dr} + \frac{d^2v}{dr^2}\right) = 0. \tag{9.18}$$

Its general solution has the form

$$v(r) = \frac{C_1}{2}r^2 + \frac{C_2}{r} + C_3 \quad \text{and} \quad v'(r) = C_1 r - \frac{C_2}{r^2}, \tag{9.19}$$

where C_j are undetermined constants. The function $u_1(\mathbf{x})$ has to be bounded at $r = 0$. Hence, in order to obtain $u_1(\mathbf{x})$ we take $C_2 = 0$ in (9.19) and the result substitutes into (9.15)

$$u_1(\mathbf{x}) = C_1 x_1, \quad |\mathbf{x}| < r_0. \tag{9.20}$$

The function $u(\mathbf{x})$ has the prescribed asymptotic at infinity (9.11). Then, (9.15) and (9.19) yield

$$u(\mathbf{x}) = x_1\left(1 + \frac{C_2}{r^3}\right), \quad |\mathbf{x}| > r_0. \tag{9.21}$$

Substitution of (9.20) and (9.21) into (9.14) gives the system of linear algebraic equations on C_1 and C_2

$$\begin{cases} C_1 = 1 + \frac{C_2}{r_0^3} \\ \lambda_1 C_1 = \lambda\left(1 - 2\frac{C_2}{r_0^3}\right). \end{cases} \tag{9.22}$$

Here, formula (9.16) is used. Solution of (9.22) is given by formulae

$$C_1 = \frac{3\lambda}{2\lambda + \lambda_1}, \quad C_2 = r_0^3\frac{\lambda - \lambda_1}{2\lambda + \lambda_1}. \tag{9.23}$$

$u_1(\mathbf{x})$ and $u(\mathbf{x})$ are given by (9.20) and (9.21), respectively. In particular,

$$u(\mathbf{x}) = x_1 \left[1 - \varrho \left(\frac{r_0}{r}\right)^3\right]. \quad |\mathbf{x}| > r_0, \tag{9.24}$$

9.2.2 Self consistent approximation

We now consider the same problem as in Sec.9.2.1 but instead of one sphere we take a large number N of small non-overlapping balls of radius r_0 with centers at the points $\mathbf{x}_k$ $(k = 1, 2, \ldots, N)$. The inclusions lie in the ball $|\mathbf{x}| < R$ where $R = \max_k |\mathbf{x}_k| + r_0$. It is assumed that the inclusions are diluted in the large ball and $r_0 \ll R$. One can consider the term $\varrho \left(\frac{r_0}{r}\right)^3$ in (9.24) as the disturbance of the plain-parallel flux (9.13) caused by a ball. Summation of all disturbances gives the total disturbance

$$M = N\varrho \left(\frac{r_0}{r}\right)^3. \tag{9.25}$$

We consider the large ball $|\mathbf{x}| < R$ as the homogenized ball of the unknown effective conductivity $\widehat{\lambda}$ embedded in a medium of conductivity λ. It follows from (9.24) that the disturbance term caused by the large ball has the form

$$M_R = \frac{\widehat{\lambda} - \lambda}{\widehat{\lambda} + 2\lambda} \left(\frac{R}{r}\right)^3. \tag{9.26}$$

Setting equal the disturbances (9.25) and (9.26) we arrive at equation

$$N\varrho \left(\frac{r_0}{r}\right)^3 = \frac{\widehat{\lambda} - \lambda}{\widehat{\lambda} + 2\lambda} \left(\frac{R}{r}\right)^3. \tag{9.27}$$

Noting that $\frac{N r_0^3}{R^3}$ is equal to the concentration ν of inclusion we obtain the fascinating Clausius-Mossotti formula for the effective conductivity (9.6)[2]. The investigation presented here is called Maxwell's approach or the self consistent approximation.

---

[2]Mossotti (1850) proposed formula (9.6) for dielectric media. Clausius (1879) discussed the same formula in the context of indices of refraction in optics. Maxwell (1873) deduced the relation (9.6) for electric conductivity in the form presented in this book in the context of heat conduction. Formula (9.6) was deduced by Lorenz (1869) and by Lorentz (1868 or 1870, not known exactly) for the refractive index in optics. It is also known as the Maxwell Garnett formula proposed for propagating waves in 1904 (for clarity, J.C. Maxwell Garnett and James Clerk Maxwell are different persons). This historic note stresses Universality of Mathematical Modelling, c.f. Table in Sec.8.7.

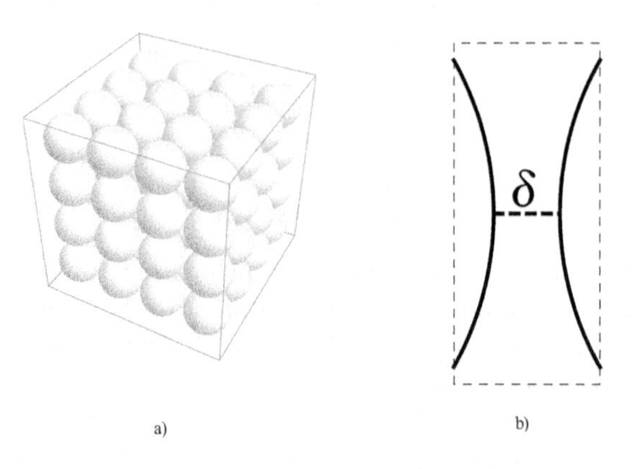

FIGURE 9.4: a) Cubic array; b) its cross section near the gap between two balls.

9.3 Densely packed balls

The Clausius-Mossotti approximation (9.6) asymptotically determines the effective conductivity of dilute balls when the concentration ν is small. In the present section, we consider another limit case when ν tends to the maximally possible value ν_c for the prescribed geometry.

9.3.1 Cubic array

Following Keller [34] we consider the cubic array of perfectly conducting balls periodically located in the space and separated by the gap $\delta = d - 2r_0$ where r_0 stands for the radius of balls and d for the distance between the neighbor balls as shown in Fig.9.4. Without loss of generality the distance d can be normalized to unity. Thus, the cubic array consists of triply periodic unit cubic cells. Asymptotic analysis will be performed for $\delta \to 0$.

Let the external field (9.12) be applied along the x_1-axis. It follows from symmetry that the heat flux goes through the cubic array separately by liens parallel to the x_1-axis. Every lien has a unit square cross section and includes an infinite chain of balls separated by the narrow gaps. Let us fix such a chain of balls. Their x_1 coordinates can be chosen as $\ell + \frac{1}{2}$ where $\ell = 0, \pm 1, \pm 2, \ldots$. The temperature in each ball is constant since the balls are filled by perfectly conducting material. It follows from symmetry that the temperature in the ℓth ball can be taken as $u_\ell = \ell$ for the corresponding flux intensity applied at infinity. Recall that the temperature can be introduced up to an arbitrary additive constant.

The main flux in the chain goes through the narrow gaps between neighbor balls. It is the same in each gap. Let us estimate the flux between the spheres centered at the points $(\pm\frac{1}{2}, 0, 0)$. The spheres are determined by equations

$$\left(x_1 \mp \frac{1}{2}\right)^2 + x_2^2 + x_3^2 = r_0^2. \tag{9.28}$$

Near the narrow gap they can be approximated by the paraboloids

$$x_1 = \pm\left(\frac{\delta}{2} + \frac{x_2^2 + x_3^2}{2r_0}\right), \quad x_2^2 + x_3^2 < r_0^2. \tag{9.29}$$

The temperature in the gap is approximated by the linear function

$$u_0(\mathbf{x}) = \frac{x_1}{H(x_2, x_3)}, \tag{9.30}$$

where $H(x_2, x_3) = \delta + \frac{x_2^2+x_3^2}{r_0}$ is the distance between the paraboloids. Then, the local flux is given by the approximate formula

$$\mathbf{q} = -\lambda \nabla u_0(\mathbf{x}) \approx -\lambda \left(\left[\delta + \frac{x_2^2 + x_3^2}{r_0}\right]^{-1}, 0, 0\right). \tag{9.31}$$

Let $B = \{(x_2, x_3) : \sqrt{x_2^2 + x_3^2} < r_0\}$ be the disk lying on the plane $x_1 = 0$. The total flux passing through the gap between two spheres can be estimated by the integral

$$g = \lambda \int_B \frac{dx_2 dx_3}{\delta + \frac{x_2^2+x_3^3}{r_0}} = \lambda \int_0^{2\pi} d\theta \int_0^{r_0} \frac{R\,dR}{\delta + \frac{R^2}{r_0}} = \lambda\pi r_0 \ln\frac{r_0}{\delta}, \tag{9.32}$$

where the double integral over B is written in polar coordinates (R, θ) with $R = \sqrt{x_2^2 + x_3^2}$ (see Sec.7.1.1 on page 135). The constant vector $(-g, 0, 0)$ can be treated as the averaged flux in the homogenzed medium, hence $(g, 0, 0) = \widehat{\lambda}\nabla x_1$. This equation yields the approximate formula for the effective conductivity for small δ

$$\frac{\widehat{\lambda}}{\lambda} \approx \pi r_0 \ln\frac{r_0}{\delta}. \tag{9.33}$$

One can see that $\widehat{\lambda}$ becomes infinite as $\delta \to 0$. Formula (9.33) can be written in terms of the concentration $\nu = \frac{4}{3}\pi r_0^3$

$$\frac{\widehat{\lambda}}{\lambda} \approx -\frac{\pi}{2}\ln\left(\frac{\pi}{6} - \nu\right), \tag{9.34}$$

where $\nu_c = \frac{\pi}{6}$ is the critical concentration of touching balls.

9.3.2 Densely packed balls and Voronoi diagrams

The Clausius-Mossotti formula (9.6) was deduced for an arbitrary macroscopically isotropic distribution of inclusions when ν is sufficiently small. Keller's formula (9.34) holds for the regular cubic array of balls when $\nu \to \frac{\pi}{6}$. In the present section, we describe the effective conductivity of densely packed balls in general case.

Let $\mathbf{e}_1 = (1,0,0)$, $\mathbf{e}_2 = (0,1,0)$, $\mathbf{e}_3 = (0,0,1,)$ be the basic vectors of $\mathbb{R}^3$ and at the same time the translation vectors generating the cubic lattice $\mathcal{Q} = \{l_i\mathbf{e}_i \equiv l_1\mathbf{e}_1 + l_2\mathbf{e}_2 + l_3\mathbf{e}_3 : \ l_j \in \mathbb{Z}\}$ where $\mathbb{Z}$ denotes the set of integer numbers[3]. In order to make a more precise statement of the problem we consider the unit cubic cell

$$Q_0 = \{\mathbf{x} = (x_1, x_2, x_3) \in \mathbb{R}^3 : \ -\frac{1}{2} < x_j < \frac{1}{2}, \ j = 1,2,3\}$$

and N points $\mathbf{x}_k \in Q_0$ $(k = 1, 2, \ldots, N)$ for which $|\mathbf{x}_k - \mathbf{x}_j| > 2r_0$ $(k \neq j)$. The location of points is triply periodically continued into the space $\mathbb{R}^3$, i.e., we have infinitely many points in the space with the centers at $\mathbf{x}_k + l_i\mathbf{e}_i$ where l_i run over the integer numbers. Consider non-overlapping perfectly conducting balls of radii r_0 centered at $\mathbf{x}_k + l_i\mathbf{e}_i$ embedded in medium of conductivity λ (see Fig.9.1). The concentration of balls holds $\nu = N\frac{4}{3}\pi r_0^3$.

It is convenient to introduce new distance (metric) in the space as follows. Two points $\mathbf{a}, \mathbf{b} \in \mathbb{R}^3$ are identified if their difference $\mathbf{a} - \mathbf{b} = l_i\mathbf{e}_i$ belongs to the lattice $\mathcal{Q}$. Hence, the topology with the opposite faces welded is introduced on Q_0[4]. In this section, the periodicity is understood as the triple periodicity with respect to the cubic lattice $\mathcal{Q}$. The distance $\|\mathbf{a} - \mathbf{b}\|$ between two points $\mathbf{a}, \mathbf{b} \in Q_0$ is introduced as

$$\|\mathbf{a} - \mathbf{b}\| := \min_{l_1, l_2, l_3 \in \mathbb{Z}} |\mathbf{a} - \mathbf{b} + l_i\mathbf{e}_i|, \tag{9.35}$$

where the modulus means the Euclidean distance in $\mathbb{R}^3$ between the points $\mathbf{a}$ and $\mathbf{b}$.

Voronoi diagrams and Delauney triangulations explain the notation of neighboring. Introduce the set of points $V = \{\mathbf{x}_k + l_i\mathbf{e}_i : k = 1, 2, \ldots, N; \ l_i \in \mathbb{Z}\}$. The Voronoi region $D(\mathbf{a})$ of a point $\mathbf{a} \in V$ is a set of all points $\mathbf{x}$ closer to $\mathbf{a}$ than to other points of V. The region $D(\mathbf{a})$ can be constructed in the following way. Let $\mathbf{b}$ be another point of V. The set of points $D_{\mathbf{b}}(\mathbf{a})$ closer to $\mathbf{a}$ than to $\mathbf{b}$ is the half-plane bisecting the segment $(\mathbf{a}, \mathbf{b})$. Then,

$$D(\mathbf{a}) = \bigcap_{\mathbf{b} \in V} D_{\mathbf{b}}(\mathbf{a}). \tag{9.36}$$

The intersection (9.36) is finite and determines a polyhedron. The Voronoi regions together with faces, edges and vertices of their boundaries form the

---

[3] c.g. the Einstein summation (7.4)

[4] In the case $\mathbb{R}^2$, the fundamental square Q_0 can be considered as the classical flat torus.

Voronoi diagram of the set V. Two Voronoi regions are called neighbor if they have a joint face. The corresponding balls are also called neighbor.

The Delauney triangulation of V (Delauney graph) consists of straight lines connecting by pairs points of V belonging to neighbor Voronoi regions[5]. It is clear that the Voronoi diagram and the Delauney graph of V are triply periodic.

We now come back to the conductivity problem. The discrete network model for densely packed balls is based on the fact that the flux is concentrated in the necks between closely spaced inclusions. For two neighbor balls B_k and B_m the relative interparticle flux g_{km} is calculated by a formula following from (9.32)

$$g_{mk} = \lambda \pi r_0 \ln \frac{r_0}{\delta_{mk}}, \tag{9.37}$$

where $\delta_{mk} = \|\mathbf{x}_m - \mathbf{x}_k\| - 2r_0$ denotes the gap between the neighbor balls in the periodic metric.

It is not necessary to take into account all the interparticle fluxes between neighbor balls since not all the gaps must be small. Hence, we can stay in the Delauney graph only edges for which $0 \leq \delta_{mk} \ll r_0$. Moreover, the term "densely packed composites" means that there exist at least three percolation chains of balls connecting the opposite faces of the cubic cell. Two percolation 2D chains are shown in Fig.9.5[6].

Let the external flux (9.12) along the x_1-axis be applied. Let u_k denote the induced constant temperature of the balls B_k and $u_k(l_i \mathbf{e}_i)$ the temperature of the ball $B_k + l_i \mathbf{e}_i$ in the cell centered at $l_i \mathbf{e}_i$ ($l_i \in \mathbb{Z}$). It follows from periodicity and symmetry of the problem that

$$u_k(l_i \mathbf{e}_i) = u_k + l_1 \quad \text{for all integer } l_i. \tag{9.38}$$

In particular, $u_k(\pm \mathbf{e}_1 + l_2 \mathbf{e}_2 + l_3 \mathbf{e}_3) = u_k \pm 1$ and $u_k(l_2 \mathbf{e}_2 + l_3 \mathbf{e}_3) = u_k$ for all integer l_2 and l_3. The heat flux is a triple periodic vector-function between inclusions. Its intensity in the gap between two neighbor balls B_k and B_j is approximated by g_{kj}. The principle of transition: *continuous* $\leftrightarrow$ *discrete* (see page 19) relates the continuous model of perfectly conducting balls and the following discrete network model.

Let the neighborhood relation between two vertexes be denoted by $\mathbf{x}_j \sim \mathbf{x}_k$ or shortly $j \sim k$. Given a triply periodic Delauney graph (V, E) where V denotes the vertex set, E the edge set. It can be reduced to the finite graph $(\mathbf{X}, E')$ where the points of $\mathbf{X} = \{\mathbf{x}_1, \mathbf{x}_2, \ldots, \mathbf{x}_N\}$ belong to the zero cubic cell Q and E' consists of the clases of periodically equivalent edges (see Fig.9.6).

---

[5] The terms *Delaunay triangulation* and *graph* used in this book are slightly different from the commonly used notations in degenerate cases [20, Chapter 10]. For simplicity consider 2D case, a square and its four vertices. The traditional 2D Delaunay triangulation has four sides of the square and one of the diagonals. In our approach, the Delaunay graph has only four sides.

[6] 2D illustrations are used in this section to simplify the presentation of complicated 3D pictures.

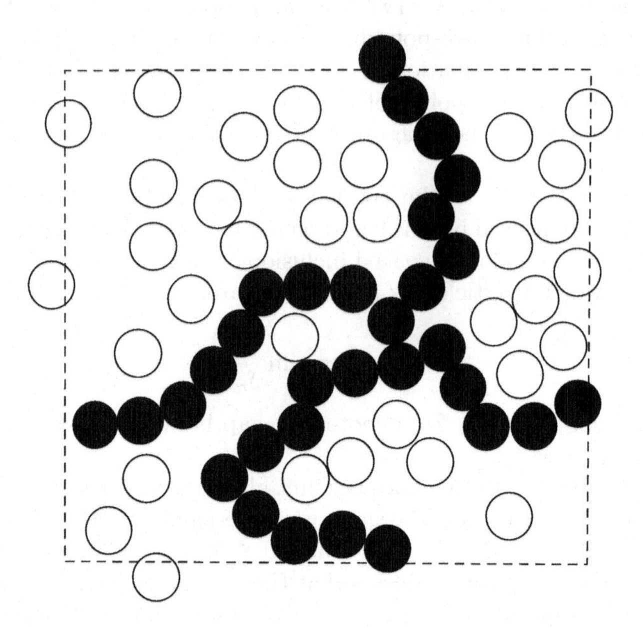

FIGURE 9.5: 2D periodic discrete network with two percolation chains connecting the opposite sides of the unit cell.

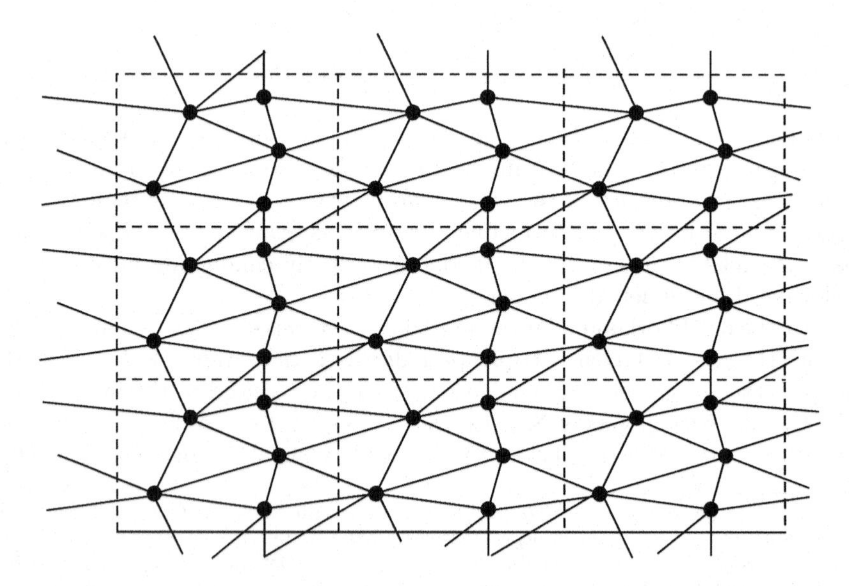

FIGURE 9.6: 2D periodic discrete network with points instead of balls. Compare to the bounded discrete network displayed in Fig.9.3.

For each fixed $\mathbf{x}_k$, introduce the set of neighbor vertices $J_k = \cup_{j\sim k}\mathbf{x}_j$. Their total number $N_k = \#J_k$ is called the degree of the vertex $\mathbf{x}_k$. A number u_k is assigned to each $\mathbf{x}_k$ and g_{mk} to each edge e_j connecting $\mathbf{x}_k$ and $\mathbf{x}_m$. The flux arisen in the network (V, E) satisfies the Kirchhoff (balance) equations

$$\sum_{j\sim k} g_{kj}(u_k - u_j^*) = 0, \quad k = 1, 2, \ldots, N, \tag{9.39}$$

where the sum $\sum_{j\sim k}$ contains N_k terms. Here, $u_j^* = u_j$ if the corresponding edge of J_k connects $\mathbf{x}_k$ with $\mathbf{x}_j \in Q_0$ and $u_j^* = u_j \pm 1$ if the edge connects $\mathbf{x}_k$ with $\mathbf{x}_j \pm \mathbf{e}_1$, respectively. Equations (9.39) can be written as the non-homogeneous system of N linear algebraic equations with N unknowns u_k

$$\sum_{j\sim k} g_{kj}(u_k - u_j) = b_k, \quad k = 1, 2, \ldots, N, \tag{9.40}$$

where b_k are determined uniquely by the graph (V, E). More precisely, $b_k = \sum_{j\sim k} g_{kj}s_{kj}$ where s_{kj} can be equal to $0, \pm 1$. It is easily seen that the sum of all equations (9.39) is a trivial equality since the terms $u_j^* = u_j - 1$ and $u_j^* = u_j + 1$ belong to (9.39) by pairs. The system of equations

$$\sum_{j\sim k} g_{kj}(u_k - u_j) = b_k, \quad k = 1, 2, \ldots, N-1, \tag{9.41}$$

with an arbitrarily fixed u_N has a unique solution. It can be proved by use of the following physical arguments. The homogeneous system (9.41) ($b_k = 0$ and $u_N = 0$) corresponds to the periodic temperature distribution when the external field is absent. This implies that the heat flux vanishes and the temperature is constant everywhere, in particular, $u_k = u_N = 0$ ($k = 1, 2, \ldots, N-1$).

All solutions of (9.40) can be treated as the set of points $\mathbf{u} = (u_1, u_2, \ldots, u_N) \in \mathbb{R}^N$ lying on a straight line parallel to the u_N-axis. Let a solution of (9.41) be known. Then, the average field energy stored in the composite is calculated as the sum of edge energies

$$E(\mathbf{u}) = \frac{\lambda}{4}\sum\nolimits^{(\mathcal{G})} g_{kj}(u_k - u_j^*)^2, \tag{9.42}$$

where the notation $\sum^{(\mathcal{G})} = \sum_{k=1}^{N}\sum_{j\sim k}$ is used for shortness. It is worth noting that this summation is determined by the fixed class $\mathcal{G}$ of graphs. The multiplier $\frac{1}{4}$ instead of usual $\frac{1}{2}$ arises due to the double counting of edges in formula (9.42). Equations (9.39) can be deduced by means of the stationary points of (9.42), i.e., by equating of all the partial derivatives $\frac{\partial E}{\partial u_k}$ to zero as it is done for mechanical systems in Sec.3.4. This corresponds to the minimization problem

$$\mathcal{E} = \min_{\mathbf{u}} E(\mathbf{u}) = \min_{\mathbf{u}} \frac{\lambda}{4}\sum\nolimits^{(\mathcal{G})} g_{kj}(u_k - u_j^*)^2. \tag{9.43}$$

The optimized function $E(\mathbf{u})$ is a positive quadratic function of several variables u_k bounded below, hence, it possesses the global minimum. This minimum is achieved on the solutions of (9.40).

The effective conductivity is determined by the macroscopic energy equation $E = \frac{1}{2}\widehat{\lambda}[u_0]_1^2$ where the increment $[u_0]_1$ of the macroscopic temperature $u_0(\mathbf{x}) = x_1$ per unit length is normalized to unity. Therefore, $E = \frac{1}{2}\widehat{\lambda}$ and

$$\widehat{\lambda} = \frac{1}{2}\sum\nolimits^{(\mathcal{G})} g_{kj}(u_k - u_j^*)^2. \tag{9.44}$$

This formula determines the effective conductivity of densely packed macroscopically isotropic balls.

9.3.3 Optimal random packing

Optimal packing of balls is attained for the face-centered cubic and the hexagonal close-packed lattices having the concentration $\nu = \frac{\pi}{3\sqrt{2}} \approx 0.74048$. Experiments show that random packing has a lower density around $0.6 - 0.7$. Optimal random close packing of balls is a geometrical problem which consists of determination of the maximally possible volume fraction ν^* of balls randomly located in the space. By random close packing of balls we understand the uniform non-overlapping distribution of balls.

On the one hand, the probabilistic uniform non-overlapping distribution of balls uniquely determines the theoretical ν_c as the mathematical expectation of the considered distribution. On the other hand, actual and computer experiments [57] demonstrate the dependence of ν_c on the protocol employed to produce the random packing. This illusory contradiction is related to unattainability of some ball locations based on the protocol restrictions. Following [57] we say that a set of contacting balls is jammed if it cannot be translated while fixing the positions of other balls. Jammed configurations are frequently met in practice. Therefore, the uniform non-overlapping distribution should be divided onto classes of structures (including jammed) having various maximally possible volume fractions.

The above stated optimal random packing problem is resolved below by introduction of the equivalence classes of graphs associated to packings. The justification of such an approach is based on the following observation [47].

Theorem 9.1. *Consider a class of macroscopically isotropic jammed balls centered at* $\mathbf{X} = (\mathbf{x}_1, \mathbf{x}_2, \cdots, \mathbf{x}_N)$ *. Let the effective conductivity* $\widehat{\lambda} = \widehat{\lambda}(\mathbf{X})$ *attain the global minimum at a location* $\mathbf{X}_*(\delta)$ *for sufficiently small* $\delta := \min_{km} \delta_{km}$*. Then, the optimal packing is attained at* $\mathbf{X}_*(0)$*.*

Proof. Let $\widehat{\lambda}(\mathbf{X})$ attain the global minimum at $\mathbf{X}_*(\delta)$ for sufficiently small δ. Let ν_* denote the concentration of balls for the location $\mathbf{X}_*(0)$. The function $\widehat{\lambda}(\mathbf{X})$ continuously depends on concentration. Hence, the function $\widehat{\lambda}(\mathbf{X}_*(\delta))$ is continuous in $0 < \delta < \delta_0$ for sufficiently small δ_0 and $\widehat{\lambda}(\mathbf{X}_*(0)) = +\infty$.

Let us assume that the optimal packing is attained at another location $\mathbf{X}^*$ for which the concentration $\nu^* > \nu_*$. The location $\mathbf{X}^*$ contains a percolation chain. Take such a radius $r_1 < r_0$ for which the concentration is reduced to ν_*. Then, all the balls in this location $\mathbf{X}^*$ with the radius r_1 are separated from each other, hence the corresponding conductivity $\widehat{\lambda}(\mathbf{X}^*)$ is a finite number. But the minimal conductivity $\widehat{\lambda}(\mathbf{X}_*(\delta))$ tends to infinity as $\delta \to 0$ for jammed balls. This yields a contradiction.

The theorem is proved.

Two graphs are called isomorphic if they contain the same number of vertices connected in the same way. Let all the periodic Delaunay graphs with N vertices be divided into the equivalence classes of isomorphic graphs. A class of graphs will be denoted by $\mathcal{G}$. We are now interested in the dependence of energy (9.42) on the spacial variable $\mathbf{X} = (\mathbf{x}_1, \mathbf{x}_2, \cdots, \mathbf{x}_N)$, i.e., $E = E(\mathbf{u}, \mathbf{X})$. Consider the minimum of the functional in a fixed class of graphs $\mathcal{G}$

$$\mathcal{E}_{\mathcal{G}} = \min_{\mathbf{u},\mathbf{X}} E(\mathbf{u},\mathbf{X}) = \min_{\mathbf{u},\mathbf{X}} \sum\nolimits^{(\mathcal{G})} f(\|\mathbf{x}_m - \mathbf{x}_k\|)(u_k - u_j^*)^2, \tag{9.45}$$

where $f(x) := \frac{\pi r_0 \lambda}{4} \ln \frac{r_0}{x - 2r_0}$. It is evident that the minimum (9.45) exists since the continuous function $f(x)$ decreases and $0 \leq f(x) \leq +\infty$ when $2r_0 \leq x \leq 3r_0$.

The function $f(x)$ as a convex function and satisfies Jensen's inequality

$$\sum_{i=1}^{M} p_i f(x_i) \geq f\left(\sum_{i=1}^{M} p_i x_i\right), \tag{9.46}$$

where the sum of positive numbers p_i is equal to unity. Equality holds if and only if all x_i are equal. Let the sum from (9.45) is arranged in such a way that $x_i = \|\mathbf{x}_k - \mathbf{x}_j\|$ and $p_i = \frac{1}{U}(u_k - u_j)^2$, where $U = \sum^{(\mathcal{G})}(u_k - u_j)^2$. Application of (9.46) to (9.45) yields

$$\sum\nolimits^{(\mathcal{G})} f(\|\mathbf{x}_m - \mathbf{x}_k\|)(u_k - u_j^*)^2 \geq U f\left(\frac{1}{U}\sum\nolimits^{(\mathcal{G})} (u_k - u_j^*)^2 \|\mathbf{x}_k - \mathbf{x}_j\|\right). \tag{9.47}$$

Hölder's inequality states that for non-negative a_i and b_i

$$\sum_{i=1}^{M} a_i b_i \leq \left(\sum_{i=1}^{M} a_i^2\right)^{\frac{1}{2}} \left(\sum_{i=1}^{M} b_i^2\right)^{\frac{1}{2}}. \tag{9.48}$$

This implies that

$$\sum\nolimits^{(\mathcal{G})} (u_k - u_j^*)^2 \|\mathbf{x}_k - \mathbf{x}_j\| \leq \left[\sum\nolimits^{(\mathcal{G})} (u_k - u_j^*)^4\right]^{\frac{1}{2}} \left[\sum\nolimits^{(\mathcal{G})} \|\mathbf{x}_k - \mathbf{x}_j\|^2\right]^{\frac{1}{2}}. \tag{9.49}$$

The function $f(x)$ decreases, hence (9.47) and (9.49) give

$$\frac{1}{U}\sum^{(\mathcal{G})} f(\|\mathbf{x}_m - \mathbf{x}_k\|)(u_k - u_j^*)^2 \geq$$
$$f\left(\frac{1}{U}\left[\sum^{(\mathcal{G})}(u_k - u_j^*)^4\right]^{\frac{1}{2}}\left[\sum^{(\mathcal{G})}\|\mathbf{x}_k - \mathbf{x}_j\|^2\right]^{\frac{1}{2}}\right). \quad (9.50)$$

The minimum of the right hand part of (9.50) on $\mathbf{X}$ is achieved independently on u_k for $\max_{\mathbf{X}} h(\mathbf{X})$ where

$$h(\mathbf{X}) = \sum^{(\mathcal{G})}\|\mathbf{x}_k - \mathbf{x}_j\|^2. \quad (9.51)$$

Lemma 9.1. *For any fixed class of graphs $\mathcal{G}$, any local maximizer of $h(\mathbf{X})$ is the global maximizer which fulfils the system of linear algebraic equations*

$$\mathbf{x}_k = \frac{1}{N_k}\sum_{j\sim k}\mathbf{x}_j + \frac{1}{N_k}\sum_{\ell=1,2,3} s_{k\ell}\mathbf{e}_\ell, \quad k = 1, 2, \ldots, N, \quad (9.52)$$

where $s_{k\ell}$ can take the values $0, \pm 1, \pm 2, \pm 3$ in accordance with the class $\mathcal{G}$. The system (9.52) has always a unique solution $\mathbf{X} = (\mathbf{x}_1, \mathbf{x}_2, \ldots, \mathbf{x}_n)$ up to an arbitrary additive constant vector.

Proof. It follows from the properties of the Voronoi tessellation that

$$\|\mathbf{x}_k - \mathbf{x}_j\| = \left|\mathbf{x}_k - \mathbf{x}_j - \sum_{\ell=1,2,3} s'_{kj\ell}\mathbf{e}_\ell\right| \quad (9.53)$$

for some $s'_{kj\ell}$ which can take the values $0, \pm 1$. The extremal points of (9.51) can be found from the system of equations

$$\nabla_k h(\mathbf{X}) = 0, \quad k = 1, 2, \ldots, N, \quad (9.54)$$

where $\mathbf{x}_k = (x_1^{(k)}, x_2^{(k)}, x_3^{(k)})$ and

$$\nabla_k = \left(\frac{\partial}{\partial x_1^{(k)}}, \frac{\partial}{\partial x_2^{(k)}}, \frac{\partial}{\partial x_3^{(k)}}\right).$$

We looking for stationary points of $h(\mathbf{X})$ in a periodic domain of $\mathbb{R}^3$ without boundary, hence all the extremal points of (9.51) satisfy this system. Equations (9.54) can be written in the equivalent form (9.52) where

$$s_{k\ell} = \sum_{j\sim k} s'_{kj\ell}.$$

One can see that the sum of all equations (9.52) gives an identity, hence, they are linearly dependent. Moreover, if $\mathbf{X} = (\mathbf{x}_1, \mathbf{x}_2, \ldots, \mathbf{x}_N)$ is a solution of (9.52), then $(\mathbf{x}_1 + \mathbf{c}, \mathbf{x}_2 + \mathbf{c}, \ldots, \mathbf{x}_N + \mathbf{c})$ is also a solution of (9.52) for any $\mathbf{c} \in \mathbb{R}^3$. Consider the homogeneous system corresponding to (9.52)

$$\mathbf{x}_k = \frac{1}{N_k} \sum_{j\ k} \mathbf{x}_j, \quad k = 1, 2, \ldots, N, \tag{9.55}$$

where $\mathbf{x}_k$ belong to the cell Q_0. The system (9.55) can be decoupled by coordinates into independent systems

$$x_p^{(k)} = \frac{1}{N_k} \sum_{j \sim k} x_p^{(j)}, \ k = 1, 2, \ldots, N \quad (p = 1, 2, 3). \tag{9.56}$$

Each pth system (9.56) has only constant solutions

$$x_p^{(1)} = x_p^{(2)} = \cdots = x_p^{(N)}. \tag{9.57}$$

This follows from the consideration of the quadratic form

$$X_p = \frac{1}{2} \sum\nolimits^{(\mathcal{G})} (x_p^{(k)} - x_p^{(j)})^2. \tag{9.58}$$

It is symmetric and positive semi-definite. Therefore, the quadratic form (9.58) has a global minimum attained at a linear set of $\mathbb{R}^N$. All the local minima of (9.56) coincide with the global minimum. One can see that the quadratic form attains the global minimum at the set (9.57). Therefore, all solutions of (9.56), hence, of (9.55) are only constants. Then, the system (9.52) has only one condition of solvability for the right hand part which is fulfilled. Therefore, the system (9.52) always has a unique solution up to an arbitrary additive constant.

The lemma is proved.

The pth coordinates $x_p^{(k)}$ of the point $\mathbf{x}_k$ ($p = 1, 2, 3$ and $k = 1, 2, \ldots, N$) are located on the real axis $\mathbb{R}$ (coincidence is permitted). For definiteness, let the Nth point be fixed as $x_p^{(N)} = 0$ ($p = 1, 2, \ldots, d$). The real numbers $x_p^{(k)}$ satisfy the system

$$x_p^{(k)} = \frac{1}{N_k} \sum_{j \sim k} x_p^{(j)} + \frac{1}{N_k} \sum_{\ell = 1,2,3} s_{kj\ell} \delta_{p\ell}, \quad k = 1, 2, \ldots, N, \tag{9.59}$$

where the Kronecker symbol $\delta_{p\ell}$ is the pth coordinate of the basic vector $\mathbf{e}_\ell$. The Nth equation (9.59) is a linear combination of the rest linearly independent equations (9.59). Solutions of (9.59) can be presented as the linear combination

$$x_p^{(k)} = \sum_{\ell = 1,2,3} x_p^{(k\ell)} \delta_{p\ell}, \quad k = 1, 2, \ldots, N, \tag{9.60}$$

where $x_p^{(k\ell)}$ satisfy the system

$$x_p^{(k\ell)} = \frac{1}{N_k} \sum_{j \sim k} x_p^{(j\ell)} + \frac{s_{k\ell}}{N_k}, \quad k = 1, 2, \ldots, N. \tag{9.61}$$

All the computations can be made separately for $p = 1, 2, 3$. As a result we obtain expressions for the optimal $\mathbf{x}_k$ in the class $\mathcal{G}$

$$\mathbf{x}_k = \sum_{\ell=1,2,3} \mathbf{x}_k^{(\ell)} \mathbf{e}_\ell, \quad k = 1, 2, \ldots, n, \tag{9.62}$$

where the vectors $\mathbf{x}_k^{(\ell)}$ can be obtained numerically.

We now proceed to summarize the algorithm to solve the optimal packing spheres problem in the periodic statement in $\mathbb{R}^3$. We consider the periodicity unit cubic cell Q_0 with welded opposite faces which contains a sufficiently large number N of spheres. The centers of spheres $\mathbf{X} = \{\mathbf{x}_1, \mathbf{x}_2, \ldots, \mathbf{x}_n\}$ lie in Q_0, satisfy the non-ovelapping restriction $\|\mathbf{x}_k - \mathbf{x}_j\| \geq 2r_0$ and the corresponding Delaunay graph Γ belong to a fixed class of graphs $\mathcal{G}$.

Let a class of graphs $\mathcal{G}$ is fixed. The Delaunay graph $\Gamma \subset \mathcal{G}$ whose vertices minimize (9.45) is called optimal. The inequalities (9.47) and (9.50) can become equalities, for instance, when some values of $(u_k - u_j^*)$ vanish and others are the same with the corresponding $\|\mathbf{x}_k - \mathbf{x}_j\|$ which are also the same. Such a case takes place for regular graphs. Then, the optimal graphs satisfy (9.52).

In the class $\mathcal{G}$, the concentration $\nu^{(\mathcal{G})}$ attains the maximal value for the optimal graph. The set of optimal graph includes graphs corresponding to packing constructed by various packing protocols. This scheme gives the set of local maximal concentrations depending on protocols, i.e., on the class of graphs. Therefore, in order to get the set of all the optimal jammed packings, it is sufficient to determine the optimal locations by means of the minimization problem (9.45). Some locations are described in Lemma 9.1.

2D example. Consider the hexagonal lattice defined by the fundamental translation vectors $\mathbf{e}_1 = \sqrt[4]{\frac{4}{3}}(1, 0)$ and $\mathbf{e}_2 = \sqrt[4]{\frac{4}{3}}(\frac{1}{2}, \frac{\sqrt{3}}{2})$. The area of the cell Q_0 holds unit. Consider $N = 3$ points $(1.075, 0.175)$, $(0.919, 0.553)$, $(0.444, 0.169)$ and the corresponding double periodic Voronoi tessellation shown in Fig.9.7a.

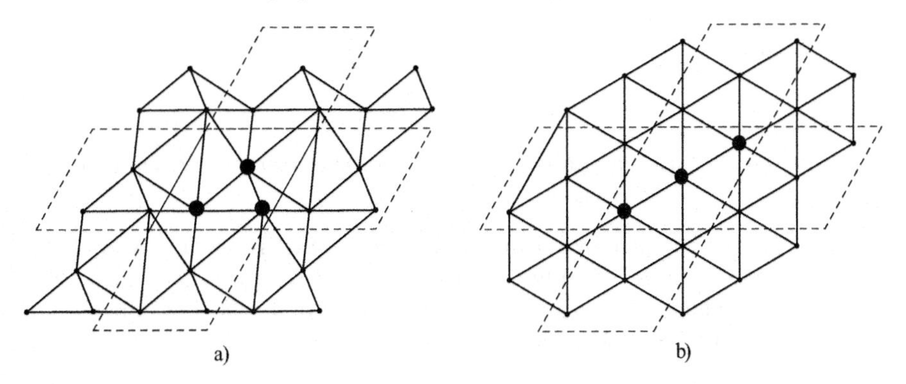

FIGURE 9.7: a) Three points in the cell Q_0 are distinguished. Dashed lines show the lattice, solid lines the double periodic Delaunay graph. b) The optimal graph isomorphic to the graph from a).

Application of the algorithm yields the optimal hexagonal structure Fig.9.7b.

Further reading. We recommend [3] for asymptotic methods and the books [1, 26, 44, 57] devoted to composites and porous media.

Exercises

1. Does formula (9.33) depend on the normalization $d = 1$ of the cubic cell?

   Hint: Check that the effective conductivity does not depend on sizes.

   Attention, it is not so for other physical processes, c.f., fluid permeability.

2. 2D problem for disks model processes in fibrous composites displayed in Fig.7.3. Deduce a formula similar to (9.6) for 2D dilute disks on the plane.

   Answer:
   $$\frac{\widehat{\lambda}}{\lambda} \approx \frac{1 + \varrho_{2D}\nu}{1 - \varrho_{2D}\nu},$$
   where $\varrho_{2D} = \frac{\lambda_1 - \lambda}{\lambda_1 + \lambda}$ denotes the 2D contrast parameter and ν the concentration of disks on the plane.

3. Deduce formulae similar to (9.33) and (9.34) for the 2D square array of disks.

Answer:

$$\frac{\widehat{\lambda}}{\lambda} \approx \pi\sqrt{\frac{r_0}{\delta}}.$$

4. Deduce a formula similar to (9.33) for the face-centered cubic and the hexagonal close-packed lattices.

5. Using the operators `VoronoiDiagram` and `DelaunayTriangulation` prepare codes for doubly and triply periodic Voronoi diagrams and Delaunay triangulations for given points. One can display both objects with `DiagramPlot` and `PlanarGraphPlot`.

6. Write and solve the system of equations (9.39) for the cubic, the face-centered cubic and the hexagonal close-packed lattices.

7. Principle by Dirac (see page 126) should be applied with care in asymptotic manipulations. Find a mistake in the following discourse.

   Transform formula (9.33) within the precision $O(\nu)$ in the case of perfectly conducting balls when $\frac{\lambda_1}{\lambda} = +\infty \Leftrightarrow \varrho = 1$:

$$\frac{\widehat{\lambda}}{\lambda} \approx \frac{1+2\nu}{1-\nu} \approx 1+3\nu \approx \frac{1}{1-3\nu}.$$

   One can see that $\frac{\widehat{\lambda}}{\lambda}$ tends to infinity as $\nu \to \frac{1}{3}$. This yields the percolation threshold $\nu_c = \frac{1}{3}$ for random composites.

   Hint 1: The following transformation is also formally valid within the precision $O(\nu)$:

$$\frac{\widehat{\lambda}}{\lambda} \approx \frac{1+2\nu}{1-\nu} \approx \frac{\nu+1}{-3\nu^3+3\nu^2-2\nu+1}.$$

   Now, the denominator has the root $\nu_c = 0.718$.

   Hint 2: Estimate the maximal concentration ν for which the above formulae give a satisfactory result, for instance, with error 10%.

8. What is the unit RVE for the hexagonal cell displayed in Fig.9.2?

   Hint: One disk, perhaps by parts, should be located in the periodicity unit cell.

9. What is wrong in the following discourse?

   Following the method of Monte Carlo (see Sec.5.3) simulate a cluster of 1000 randomly distributed non-overlapping balls in $\mathbb{R}^3$ with a high local concentration. Let us have in our disposal Laplace's device[7] *(supercode*

[7] "Give me the positions and velocities of all the particles in the universe, and I will predict the future." - Marquis Pierre Simon de Laplace.
By the way, one can estimate the time of computations or the size of computer to perform Laplace's task, c.f., Principle of the energy estimations on page 16.

like FEM, supercomputer etc.) and be able to compute the local fields. After substitution of the computed fields into (9.3) *we arrive at a formula for the effective conductivity of densely packed balls.*

Hint: The volume fraction of 1000 balls in $\mathbb{R}^3$ is equal to zero.

Actually, a formula deduced in such a way gives the effective conductivity of dilute oriented clusters in $\mathbb{R}^3$. Every cluster consists of 1000 balls.

10. In order to prove that the minimization problem (9.43) possesses the global minimum, first, check that it has local minima. After, check that all the local minima coincide with the global minimum.

11. Fix the number of balls per cell, for instance $N = 1$, $N = 2$, ... and a class of graphs. Construct the optimal graph in the considered class of graphs following the example given on page 212.

12. Consider a random set $\mathbf{X} = \{\mathbf{x}_1, \mathbf{x}_2, \ldots, \mathbf{x}_N\}$ of points in the unit cubic cell or in the unit square and their random walks under restrictions $\|\mathbf{x}_k - \mathbf{x}_j\| \geq r_0$. Construct the corresponding Delaunay triangulations moving simultaneously with the random walks. Select the statistic data of the moving Delaunay triangulations (number of edges, degrees of vertices). Investigate the critical values of N and r_0 when the random walks do not change the class of graphs.

13. Prepare a code to automatically transform a picture with two-phase medium onto the data $\{(\mathbf{x}_k, d_k)\}_{k=1}^N$ where $\mathbf{x}_k$ denotes the coordinates of a inner point of an inclusion, d_k its characteristic diameter. Construct the corresponding Delaunay graph.

Bibliography

[1] Adler, P.M., Thovert, J.-F., Mourzenko, V.V.: *Fractured porous media.* Oxford University Press, Oxford (2012)

[2] Adler, P., Malevich, A.E., Mityushev, V.: Nonlinear correction to Darcy's law. *Acta Mechanica* **224**, 1823–1848 (2013)

[3] Andrianov, I.V., Manevitch, L.I.: *Asymptotology: Ideas, Methods, and Applications.* Kluwer Academic Publishers, Dordrecht etc (2002)

[4] Antosik, P., Mikusiński, J, Sikorski, R.: *Theory of distributions, The sequential approach.* Elsevier - PWN, Amsterdam - Warszawa (1973)

[5] Amelkin, V.V.: *Differential Equations in Applications.* Mir Publ., Moscow (1990)

[6] Arnold, V.I., Khesin, B.A.: *Topological Methods in Hydrodynamics.* Springer, New York (1998)

[7] Arnold, V. I.: *Catastrophe Theory, 3rd ed.* Berlin, Springer-Verlag (1992).

[8] Arnold, V.I.: On teaching mathematics. *Uspekhi Mat. Nauk* **53**, 229–234 (1998); English translation: *Russian Math. Surveys* **53**, 229–236 (1998)

[9] Artstein, Z.: *Mathematics and the Real World: The Remarkable Role of Evolution in the Making of Mathematics.* Prometheus Books, Amherst New York (2014)

[10] Batchelor, G.K.: Transport properties of two-phase materials with random structure. *Annual Review of Fluid Mechanics* **6**, 227–255 (1974)

[11] Berlyand, L., Kolpakov, A.G., Novikov, A.: *Introduction to the Network Approximation Method for Materials Modeling.* Cambridge University Press;, Cambridge (2012)

[12] Berrut, J.-P., Trefethen, L.N.: Barycentric Lagrange Interpolation. *SIAM Review* **46**, 501–517 (2004)

[13] Borisenko, A.I., Tarapov, I.E.: *Vector and Tensor Analysis with Applications.* Prentice-Hall, Englewood Cliffs (1968)

[14] Buchberger B. et al.: Symbolic Computation (An Editorial), *Journal of Symbolic Computation*, 1, 16, 1985.

[15] Bungartz, H.-J., Zimmer, S., Buchholz, M., Pflüger, D.: *Modeling and Simulation. An Application-Oriented Introduction*, Springer-Verlag, New York etc (2014)

[16] Courant, R., Hilbert, D.: *Methods of Mathematical Physics.* V. 1–2. Wiley-Interscience, Weinheim (1989)

[17] Cohen J. S.: *Computer algebra and symbolic computations : elementary algorithms.* A K Peters, Ltd., 2002.

[18] Cohen J. S.: *Computer algebra and symbolic computations : mathematical methods.* A K Peters, Ltd., 2003.

[19] Esfandiari R.S.: *Numerical Methods for Engineers and Scientists Using MATLAB®*. Chapman and Hall/CRC, NY (2017)

[20] Gallier, J.: *Geometric Methods and Applications.* Springer-Verlag, New York etc (2011)

[21] Gelfand, I.M., Shilov, G.E.: *Generalized functions.* V. 1–5, Academic Press, New York (1966–1968)

[22] Ghoniem, N., Walgraef, D.: *Instabilities and Self-Organization in Materials.* V. 1–2, Oxford Univ. Press, Oxford (2008)

[23] Grzymkowski, R., Słota, D.: *Computational Methods for Integral Equations.* Śląsk Technological University Publ., Gliwice (2015) [in Polish]

[24] Falconer, K.: *Fractals, A Very Short Introduction.* Oxford University Press, Oxford (2013)

[25] Faul A.C.: *A Concise Introduction to Numerical Analysis.* Chapman and Hall/CRC, NY (2016)

[26] Gluzman, S., Mityushev, V., Nawalaniec, W.: *Computational Analysis of Structured Media.* Elsevier, Amsterdam (2017)

[27] Guillermo J., León S.: *Mathematica Beyond Mathematics: The Wolfram Language in the Real World.* Chapman and Hall/CRC, NY (2017)

[28] Hazrat R. *Mathematica: A Problem-Centered Approach.* Springer (2015)

[29] Hahn B., Valentine D.T.: *Essential MATLAB for Engineers and Scientists.* Academic Press (2016).

[30] Hastings C., Mischo K., Morrison M.: *Hands-On Start to Wolfram Mathematica: And Programming with the Wolfram Language.* Wolfram Media (2016).

[31] Howison, S.: *Practical Applied Mathematics: Modelling, Analysis, Approximation.* Cambridge University Press, Cambridge (2004)

[32] Illner, R., Bohum, C.S., McCollum, S., van Roode, Th.: *Mathematical Modeling: A Case Studies Approach.* AMS, Rhode Island (2005)

[33] Jost, J: *Mathematical Methods in Biology and Neurobiology.* Springer-Verlag, London (2014)

[34] Keller, J.B.: Conductivity of a Medium Containing a Dense Array of Perfectly Conducting Spheres or Cylinders or Nonconducting Cylinders. *J. Appl. Phys.* **34**, 991-993 (1963)

[35] Kincaid, D, Cheney, W.: *Numerical Analysis. Mathematics of Scientific Computing.* 3rd edition. AMS, Rhode Island (2009)

[36] Kolpakov, A.A., Kolpakov, A.G.: *Capacity and Transport in Contrast Composite Structures: Asymptotic Analysis and Applications.* CRC Press Inc., Boca Raton etc (2009)

[37] Lawler, G.F.: *Random Walk and the Heat Equation.* AMS, Rhode Island (2010)

[38] Levi, M.: *The Mathematical Mechanic: Using Physical Reasoning to Solve Problems.* Princeton University Press, Princeton and Oxford (2009)

[39] Levi, M.: *Classical Mechanics with Calculus of Variations and Optimal Control: An Intuitive Introduction.* AMS, Rhode Island (2009)

[40] Levi, M.: *Why Cats Land on Their Feet: And 76 Other Physical Paradoxes and Puzzles.* Princeton University Press, Princeton (2012)

[41] Logan, J.D.: *Applied Mathematics.* Wiley, 4th Edition, Hoboken (2013)

[42] Lykov, A.B.: *Heat and Mass Transfer.* Mir Publ., Moscow (1980) [in Russian]

[43] Mangano, S.: *Mathematica Cookbook.* O'Reilly Media, Sebastopol (2010)

[44] Milton, G.W.: *The Theory of Composites*, Cambridge University Press, Cambridge (2012)

[45] Maxwell, J. C., *A Treatise on Electricity and Magnetism*, Dover, New York (1954). (Republication of 3rd edition of 1891, 1st edition 1873.)

[46] Mityushev, V., Nawalaniec, W., Rylko, N., Malevich, A.: *Computational Method of Industrial Mathematics.* V. 1–3. Jacek Skalmierski Publ., Gliwice (2010); Open access `mityu.up.krakow.pl/publication/`

[47] Mityushev V., Rylko N.: Optimal distribution of the non-overlapping conducting disks. *Multiscale Model. Simul.* **10**, 180-190 (2012)

[48] Murray, J.D.: *Mathematical Biology.* Springer-Verlag, Berlin Heidelberg v.1 (2002); v.2 (2003)

[49] Myshkis, A.D.: *Elements of the theory of mathematical models.* URSS Publ., Moscow (2007) [in Russian]

[50] Paule P., Kartashova L., Kauers M., Schneider C., Winkler F.: Hot Topics in Symbolic Computation. In B. Buchberger et al. (Eds.), *Hagenberg Research*, Springer, 2009.

[51] Pinkus, A.: Approximation theory of the MLP model in neural net works, *Acta Numerica* **8**, 143-195 (1999)

[52] Pinsky, M.A.: *Partial Differential Equations and Boundary-Value Problems with Applications.* Third Edition. AMS, Rhode Island (1998)

[53] Rylko, N.: Representative volume element in 2D for disks and in 3D for balls. *Journal of Mechanics of Materials and Structures* **9**, 427-439 (2014)

[54] Schwartz, L.: *Théorie des distributions.* V. 1-2, Hermann, Paris (1950-1951)

[55] Stakgold, I.: *Green's Functions and Boundary Value Problems.* Wiley, Hoboken (1998)

[56] Stormy, I.: *Matlab. A Practical Introduction to Programming and Problem Solving.* Butterworth-Heinemann (2016)

[57] Torquato, S.: *Random Heterogeneous Materials: Microstructure and Macroscopic Properties.* New York, Springer-Verlag (2002)

[58] Verzani J.: *Using R for Introductory Statistics.* Chapman & Hall/CRC, Boca Raton, Florida (2005)

[59] Von Zur Gathen J., Gerhard J.: *Modern Computer Algebra.* Cambridge University Press, 3rd Edition (2013)

[60] Wolfram S.: *An Elementary Introduction to the Wolfram Language.* Wolfram Media (2017)

[61] Zeldovich, Ja.B., Myshkis, A.D.: *Elements of mathematical physics. Medium consisting of non-interacting particles.* Nauka, Moscow (1973) [in Russian]

Index

1D, 45
2D, 149
3D, 113

amplitude, 75
angle velocity, 143

Banach Fixed Point Theorem, 34
bifurcation point, 66
black box, 20

Cauchy's problem, 5, 35
convolution, 20
correlation coefficient, 96
curl, rot, 156
cybernetics, 20
cylindrical coordinates, 135

damped motion, 76
diffusion constant, 107
Dirac δ-function, 128
direction cosines, 134
Dirichlet boundary condition in 1D, 118
Dirichlet boundary value problem, 168
Dirichlet function, 19
divergence, div, 152

eigenvalue problem, 87
energy of particle, 54
equilibrium point, 87
equilibrium state, 64

Financial data, 109
fitting, fit, 97
Fourier coefficients, 173
Fourier law, 117
Fourier law for anisotropic media, 164
Fourier law for isotropic media, 164
Fourier series, 173
frequency, 75

Gauss's formula, see Ostrogradsky-Gauss's formula, 160
gradient, operator ∇, 148
Green's function in 1D, 122

harmonic oscillation (motion), 73
heat capacity, 165
heat conduction coefficient, 116
heat equation, 166
heat flux, 116
Heaviside's function, 118
Hooke's law, 20

inertia tensor, 146
initial phase, 75
integral operator, 20
interpolation, 46
Iteration, 29
Iterative method, 29

Jacobi matrix, 87

Kepler's equation, 38
kinetic energy, 76

least squares method (LSM), 94
level curve, 150
level surface, 152
linear operator, 19
linearization, 86
logistic differential equation, 61

Lotka-Volterra model, 83

Malthus population law, 60
mechanical meaning of the derivative, 45
mixed product, 141
moment of a force, 140
Monte Carlo method, 102

natural frequency, 81
natural parameter on curve, 149
Neumann boundary value problem, 168
Newton's method, 30
normal vector, 149

Ostrogradsky-Gauss's formula, 160

Padé approximation, 71
Parseval's identity, 174
phase space, 55
Pi theorem, 24
polar coordinates, 135
potential energy, 54
predator-prey model, 83
principle by Arnold, 159
principle by Dirac, 126
principle of an apple, 100
principle of Asymptotology, 193
principle of hand calculations, 6
principle of innocence, 10
principle of microscope, 38
principle of stability, 22
principle of stupid computer, 10
principle of the balanced computations, 25
principle of the energy estimations, 16
principle of the simplest model, 12
principle of transition:
continuous $\leftrightarrow$ *discrete*, 19
principle of trying, 124
principle of units, 23
projectile, 4

radioactive decay, 59
regression equation, 94
resonance, 81

scalar field, 148
scalar product, 138
separation of variables, 177
spherical coordinates, 138
stability of the mathematical model, 21
standard deviation, 96
Stokes's formula, 160

Taylor formula, 69
temperature distribution, 116
tensor of order 1, 146
tensor of order 2, 146
total energy, 76
trigonometric approximation, 99
trigonometric series, 173

vector field, 148
vector product, 139